Assessing Computational Thinking

This book presents different approaches for answering the question: How do we assess computational thinking? The result is a snapshot of the current state of the field for assessing computational thinking.

The last decade has seen rapid growth in the presence of computational thinking (CT) in educational contexts. Those working to advance CT argue that the concepts and skills associated with CT are essential to succeed in an increasingly computational world. As a result of these efforts, there has been tremendous growth in curricula, learning environments, and innovations around CT education in K-12 classrooms and beyond. As CT grows in prominence, so too does the need to be able to effectively and equitably assess learners CT abilities. This volume is a collection of chapters pursuing different approaches for answering the question: How do we assess computational thinking? The answers provided span age ranges, formal and informal contexts, conceptual aspects of CT, and varying methodological and evaluative strategies. Collectively, this volume captures the current state of the field for assessing computational thinking and lays the groundwork for future CT assessment innovation.

Assessing Computational Thinking will be a key resource for academics, researchers, and advanced students of Education, Educational Assessment, Educational Research, Psychology, and Research Methods. The chapters included in this book were originally published as a special issue of *Computer Science Education*.

David Weintrop is Assistant Professor at the University of Maryland, College Park, USA.

Daisy W. Rutstein is Principal Education Researcher at SRI International, Menlo Park, USA.

Marie Bienkowski is Director of STEM Curriculum at the Academies of Math and Science, USA.

Steven McGee is President of The Learning Partnership, USA.

Assessing Computational Thinking

An Overview of the Field

Edited by

David Weintrop, Daisy Wise Rutstein, Marie Bienkowski and Steven McGee

Routledge
Taylor & Francis Group
LONDON AND NEW YORK

First published 2024
by Routledge
4 Park Square, Milton Park, Abingdon, Oxon, OX14 4RN

and by Routledge
605 Third Avenue, New York, NY 10158

Routledge is an imprint of the Taylor & Francis Group, an informa business

British Library Cataloguing-in-Publication Data
A catalogue record for this book is available from the British Library

ISBN13: 978-1-032-55530-0 (hbk)
ISBN13: 978-1-032-55531-7 (pbk)
ISBN13: 978-1-003-43115-2 (ebk)

DOI: 10.4324/9781003431152

Typeset in Myriad Pro
by codeMantra

Publisher's Note
The publisher accepts responsibility for any inconsistencies that may have arisen during the conversion of this book from journal articles to book chapters, namely the inclusion of journal terminology.

Disclaimer
Every effort has been made to contact copyright holders for their permission to reprint material in this book. The publishers would be grateful to hear from any copyright holder who is not here acknowledged and will undertake to rectify any errors or omissions in future editions of this book.

Contents

Citation Information

The chapters in this book were originally published in the journal *Computer Science Education*, volume 31, issue 2 (2021). When citing this material, please use the original page numbering for each article, as follows:

Introduction

Assessing computational thinking: an overview of the field
David Weintrop, Daisy Wise Rutstein, Marie Bienkowski and Steven McGee
Computer Science Education, volume 31, issue 2 (2021) pp. 113–116

Chapter 1

Developing a kindergarten computational thinking assessment using evidence-centered design: the case of algorithmic thinking
Jody Clarke-Midura, Deborah Silvis, Jessica F. Shumway, Victor R. Lee and Joseph S. Kozlowski
Computer Science Education, volume 31, issue 2 (2021) pp. 117–140

Chapter 2

Design and validation of learning trajectory-based assessments for computational thinking in upper elementary grades
Brian D. Gane, Maya Israel, Noor Elagha, Wei Yan, Feiya Luo and James W. Pellegrino
Computer Science Education, volume 31, issue 2 (2021) pp. 141–168

Chapter 3

A principled approach to designing computational thinking concepts and practices assessments for upper elementary grades
Satabdi Basu, Daisy W. Rutstein, Yuning Xu, Haiwen Wang and Linda Shear
Computer Science Education, volume 31, issue 2 (2021) pp. 169–198

Chapter 4

Assessing computational thinking through the lenses of functionality and computational fluency
Shari J. Metcalf, Joseph M. Reilly, Soobin Jeon, Annie Wang, Allyson Pyers, Karen Brennan and Chris Dede
Computer Science Education, volume 31, issue 2 (2021) pp. 199–223

Chapter 5

Communicating about computational thinking: understanding affordances of portfolios for assessing high school students' computational thinking and participation practices
Deborah Fields, Debora Lui, Yasmin Kafai, Gayithri Jayathirtha, Justice Walker and Mia Shaw
Computer Science Education, volume 31, issue 2 (2021) pp. 224–258

Chapter 6

How do students develop computational thinking? Assessing early programmers in a maze-based online game
Mariluz Guenaga, Andoni Eguíluz, Pablo Garaizar and Juanjo Gibaja
Computer Science Education, volume 31, issue 2 (2021) pp. 259–289

Chapter 7

Assessing computational thinking in libraries
David Weintrop, Shandra Morehouse and Mega Subramaniam
Computer Science Education, volume 31, issue 2 (2021) pp. 290–311

Notes on Contributors

Satabdi Basu, Center for Education Research & Innovation, SRI International, Menlo Park, USA

Marie Bienkowski, Academies of Math and Science, USA.

Karen Brennan, Harvard Graduate School of Education, Harvard University, Cambridge, USA.

Jody Clarke-Midura, Instructional Technology and Learning Science, Utah State University, Logan, USA.

Chris Dede, Harvard Graduate School of Education, Harvard University, Cambridge, USA.

Andoni Eguíluz, Engineering Faculty, University of Deusto, Bilbao, Spain.

Noor Elagha, Psychology, University of Illinois at Chicago, USA.

Deborah Fields, Instructional Technologies and Learning Science, Utah State University, Logan, USA.

Brian D. Gane, Learning Sciences Research Institute, UIC's College of Liberal Arts and Sciences, Chicago, USA.

Pablo Garaizar, Engineering Faculty, University of Deusto, Bilbao, Spain.

Juanjo Gibaja, Deusto Business School, University of Deusto, Bilbao, Spain.

Mariluz Guenaga, Engineering Faculty, University of Deusto, Bilbao, Spain.

Maya Israel, Educational Technology, University of Florida, Gainesville, USA.

Gayithri Jayathirtha, University of Pennsylvania, Philadelphia, USA.

Soobin Jeon, Harvard Graduate School of Education, Harvard University, Cambridge, USA.

Yasmin Kafai, Instructional Technologies and Learning Science, University of Pennsylvania, Philadelphia, USA.

Joseph S. Kozlowski, College of Teacher Education and Leadership, Utah State University, Logan, USA.

Victor R. Lee, Graduate School of Teacher Education, Stanford University, USA.

Debora Lui, Computer Science and Engineering, The Baldwin School, USA.

Feiya Luo, Educational Technology, University of Florida, Gainesville, USA.

Shari J. Metcalf, Harvard Graduate School of Education, Harvard University, Cambridge, USA.

Steven McGee, The Learning Partnership, USA.

Shandra Morehouse, College of Information Studies, University of Maryland, College Park, USA.

James W. Pellegrino, Learning Sciences Research Institute, UIC's College of Liberal Arts and Sciences, Chicago, USA.

Allyson Pyers, Harvard Graduate School of Education, Harvard University, Cambridge, USA.

Joseph M. Reilly, Harvard Graduate School of Education, Harvard University, Cambridge, USA.

Daisy W. Rutstein, Center for Education Research & Innovation, SRI International, Menlo Park, USA.

Mia Shaw, University of Pennsylvania, Philadelphia, USA.

Linda Shear, Center for Education Research & Innovation, SRI International, Menlo Park, USA.

Jessica F. Shumway, College of Teacher Education and Leadership, Utah State University, Logan, USA.

Deborah Silvis, Instructional Technology and Learning Science, Utah State University, Logan, USA.

Mega Subramaniam, College of Information Studies, University of Maryland, College Park, USA.

Justice Walker, University of Pennsylvania, Philadelphia, USA; University of Texas, El Paso, USA.

Annie Wang, Harvard Graduate School of Education, Harvard University, Cambridge, USA.

Haiwen Wang, Center for Education Research & Innovation, SRI International, Menlo Park, USA.

David Weintrop, College of Education & College of Information Studies, University of Maryland, College Park, USA.

Yuning Xu, Kidaptive Inc, USA.

Wei Yan, Educational Technology, University of Florida, Gainesville, USA.

INTRODUCTION

Assessing computational thinking: an overview of the field

The last decade has seen rapid growth in the presence of computational thinking (CT) in educational contexts. Those working to advance CT argue that the concepts and skills associated with CT are essential to succeed in an increasingly computational world. As a result of these efforts, CT has a growing presence in K-12 classrooms and beyond. This can be seen in the inclusion of CT in disciplinary standards (e.g. the Next Generation Science Standards and Common Core Math identifying CT as a core practice), as well as national curricular efforts (e.g. the United Kingdom's national computing curriculum seeks to have students "develop and apply their analytic, problem-solving, design, and computational thinking skills"). Just as CT has a growing presence in formal education, it can also be seen in informal contexts through the growth of computing camps, after-school and library CT programming, and a growing array of toys designed to engage youth with CT.

The contemporary discussion around CT began with Wing's (2006) article, where she argued "to reading, writing, and arithmetic, we should add computational thinking to every child's analytical ability" (p. 33). However, the conceptual origins have a much longer history, dating back to early work on the Logo programming language and Papert's insights on the potential of computing as a powerful context for learning (1980). In response to Wing's article, much effort has been dedicated to trying to define what constitutes CT and where the boundaries of the construct lie. While the community has yet to settle on a single unified definition, there is general consensus that CT includes foundational computing concepts such as abstraction and algorithms, as well as computing practices such as problem decomposition and debugging (Grover & Pea, 2013; Shute et al., 2017). As the dust started to settle from early debates around the scope and nature of CT, a growing number of research projects sought to design CT learning experiences. Spurred in part by an increase in funding for educational projects at the intersection of computing and other disciplines, a space in which CT is particularly well-suited to contribute, the last decade has seen tremendous growth in curricula, learning environments, and innovations around CT education (Tang et al., 2020). In the wake of this growth, this special issue seeks to respond to a question of growing importance: *How do we assess computational thinking*?

This is not a straightforward question to answer as several aspects of CT make it challenging to assess. For example, there is a wide variety of methods by which CT is taught and contexts in which students learn CT. While some schools offer stand-alone CT learning experiences, other schools may try to integrate CT within current subject matters. Further, as discussed above, CT is a relatively ill-defined construct, thus, different assessments may focus on slightly different dimensions of CT. Collectively, this produces a landscape where a variety of assessments are needed to reflect the different conceptual, contextual, and motivational aspects of CT instruction.

In selecting work for this special issue, a concerted effort was made to capture the plurality of ways in which CT is being assessed, while at the same time demonstrating the importance of building upon the fundamentals of assessment design. Towards this end, submissions were categorized based on several dimensions, including learning context (e.g. kindergarten classroom), age/grade of learners, CT definition, relation to other disciplines (e.g. integrated vs. stand-alone), methodological approach, and assessment type. The result is a set of seven papers that span these categories. We briefly summarize this work below, providing a picture of the breadth of research on assessing CT, beginning with work focused on young learners.

As the presence of CT in early elementary classrooms grows, so too does the need to have effective assessments for such early learning contexts. Towards this end, Clarke-Midura and colleagues document the development of an empirically grounded assessment in their paper entitled *Developing a kindergarten computational thinking assessment using evidence-centered design: the case of algorithmic thinking*. This work is novel both in the context (CT in kindergarten) as well as in the approach to assessing CT in a screen-free context, relying on paper-based activities to evaluate young learners' emerging algorithmic thinking skills. This work expands the landscape of CT assessment to consider the assessment of CT in unplugged contexts with young children.

Continuing to slightly older learners, Gane and colleagues draw on learning trajectories to develop an assessment for CT situated in a mathematics context. Their paper, entitled *Design and validation of learning trajectory-based assessments for computational thinking in upper elementary grades*, focuses on assessing learners in grades 3 and 4 on the constructs of sequence, repetition, conditionals, decomposition, variables, and debugging. This work demonstrates how learning, instruction, and assessment can be coordinated based on a model of what it means to learn and know CT.

Not all learning objectives can be measured using the same type of tasks or assessment approach, a fact that must be taken into account during the design process. This is especially true in the context of CT given its various facets, which can make designing CT assessments particularly difficult. Basu and colleagues speak directly to this challenge in their paper titled *A principled approach to designing computational thinking concepts and practice for upper elementary grades*. They show how an evidence-centered design approach can be used to develop separate assessments for measuring concepts and practices using easily-scorable tasks. The paper discusses how a set of focal knowledge, skills, and abilities were developed that matched the purpose and requirements of the assessment. The paper further discusses the design process and provides results of the administration of the two assessments as part of a large-scale evaluation project.

Shifting away from written assessment, Metcalf and colleagues investigate assessments that ask learners to enact CT practices, employing their CT knowledge and skills to generate computational artifacts. In their paper *Assessing computational thinking through the lenses of functionality and computational fluency*, Metcalf and colleagues present two rubrics that can be applied to assess student-created artifacts, one that focused on the functionality of the resulting program and the other that focused on the stages of development. These two rubrics reflect the desire to assess both declarative and procedural CT competency.

In response to a recognition of the multifaceted nature of CT, Fields and colleagues report on research using portfolio analysis to identify and assess students' ability to communicate about computational practices in their paper titled *Communicating about Computational Thinking: Understanding Affordances of Portfolios for Assessing High School Students' Computational Thinking and Participation Practices*. This work focuses on a high-school unit in which students work on e-textile projects and create portfolios containing the artifacts they create alongside a description of their process and the challenges they encountered along the way. This work contributes along several novel dimensions to the picture of CT assessment painted by this special issue, namely another methodological approach (portfolio review), a novel CT learning context (e-textiles in high school classrooms), and additional dimensions of CT (communication).

One feature of many CT learning experiences situated in computational contexts is the ability to collect fine-grained data on student performance by logging student interactions. In their paper entitled *How Do Students Develop Computational Thinking? Challenges and Difficulties Assessing Early Programmers in a Maze-Based Online Game*, Guenaga and colleagues look at data from young students (ages 8–14) who used an instrumented coding platform to solve coding puzzles. Drawing on best practices from learning analytics research, this work provides useful tools and techniques that can be used to create a more well-rounded picture of student's knowledge. Such assessments provide insight into students' understanding of CT concepts as well as ways the assessment itself, in this case the maze-based puzzles, can be improved.

Acknowledging the plurality of context where CT learning occurs, this special issue also includes work by Weintrop and colleagues exploring how CT is assessed in library contexts. In their paper, entitled *Assessing Computational Thinking in Libraries*, Weintrop and colleagues interviewed 37 library staff members from across the United States to identify the motivations library staff members have for bringing CT into their buildings, the challenges they face in trying to assess CT, and what library would like to assess about their CT programming. This work contributes new insights into the motivations, challenges, and aspirations of those providing CT learning experiences beyond the walls of the classroom.

Looking across the research compiled for this special issue, this collection provides insight into the different ways in which CT is being assessed and the methods used for creating innovative CT assessments. As the field of CT continues to grow, so to must the landscape of CT assessments. While we know there are many other projects and assessments beyond what is captured here, we hope that these articles will inspire new tools, methods, and insights that advance the goal of creating assessments that live up to the lofty aspirations of CT.

Funding Acknowledgement

Steven McGee was supported in part by National Science Foundation grants CNS-1640215, CNS-1738572 to The Learning Partnership. Any opinions, findings, and conclusions or recommendations expressed in this material are those of the authors and do not necessarily reflect the views of NSF.

References

Grover, S., & Pea, R. (2013). Computational thinking in K-12: A review of the state of the field. *Educational Researcher*, *42*(1), 38–43. https://doi.org/10.3102/0013189X12463051

Papert, S. (1980). *Mindstorms: Children, computers, and powerful ideas*. Basic books.

Shute, V. J., Sun, C., & Asbell-Clarke, J. (2017). Demystifying computational thinking. *Educational Research Review, 22*, 142–158. https://doi.org/10.1016/j.edurev.2017.09.003

Tang, X., Yin, Y., Lin, Q., Hadad, R., & Zhai, X. (2020). Assessing computational thinking: A systematic review of empirical studies. *Computers & Education, 148*, 103798. https://doi.org/10.1016/j.compedu.2019.103798

Wing, J. M. (2006). Computational thinking. *Communications of the ACM, 49*(3), 33–35. https://doi.org/10.1145/1118178.1118215

David Weintrop

Daisy W. Rutstein

Marie Bienkowski

Steven McGee

Developing a kindergarten computational thinking assessment using evidence-centered design: the case of algorithmic thinking

Jody Clarke-Midura, Deborah Silvis, Jessica F. Shumway, Victor R. Lee and Joseph S. Kozlowski

ABSTRACT

Background and Context: There is a need for early childhood assessments of computational thinking (CT). However, there is not consensus on a guiding framework, definition, or set of proxies in which to measure CT. We are addressing this problem by using Evidence Centered Design (ECD) to develop an assessment of kindergarten-aged children's CT.

Objective: To present a design case on the development of the assessment, specifically the algorithmic thinking (AT) tasks and to share validity evidence that emerged.

Method: We focus on the AT sub-component of CT and present the principled assessment design process using ECD.

Findings: Our operationalization of CT includes spatial reasoning as a sub-component. Pilot results showed an acceptable internal consistency reliability for the AT items and critical design decisions that contributed to validity evidence.

Implications: An important contribution of this work is the inclusion of spatial reasoning in our definition of early childhood CT.

Introduction

Early childhood is increasingly seen as a time to introduce and foster computer science skills and computational thinking (CT). While this trend is not without historical precedent (e.g., Papert, 1980), there is currently a movement to produce computationally-themed materials and toys for kindergarten-aged children (Hamilton et al., 2020). In some cases, the design of these materials and toys are informed by research (e.g., Bers et al., 2014). In many other cases, the use and development of such materials and toys are driven by popular intuition. As the whole range of materials are beginning to see uptake in kindergarten and early childhood settings, we see a more pressing need for there to be assessments for this age group that measure CT.

However, for a number of reasons, determining what to assess and how to do so is not as simple as it seems (Pellegrino et al., 2016). First, the field does not currently have an agreed upon definition of CT. Unlike assessments in disciplines like math (Clements et al., 2019), there is not yet consensus on what skills constitute CT. Second, even if an agreed-upon definition of CT was established, we still lack a clear understanding of the knowledge, skills, and abilities (KSAs) involved in young children's CT given some developmental constraints. For example, Kindergarten children are just learning how to read and write, meaning the expression of their CT capabilities do not lend themselves to many standard forms of computer-based or paper-and-pencil elicitation techniques. Third, and relatedly, interest in early childhood CT is so recent we also lack common curriculum practices to observe that could inform assessment design. For these reasons and more, it is easy to come unmoored because so much is unknown and in need of study. In these messy circumstances, it helps to put a stake in something, and for us that was the Evidence-Centered Design process (ECD, Mislevy & Haertel, 2006).

The present article applies ECD to the early stages of the design of an assessment of kindergarteners' CT. Specifically, our goal has been to design an assessment that can be used to assess CT as promoted in currently designed materials and toys for young children who are preliterate, which emphasize controlling and moving an agent through a grid-like space. Aligning our design process with a rigorous set of established steps for setting up our assessment as an argument through conducting domain analysis, domain modeling, and conceptual assessment framing, provided structure for assessment design. This approach supported making valid inferences about children's learning, based on iteratively specifying and testing their CT knowledge, skills, and abilities. ECD was a means of managing the uniquely messy circumstances of assessing young children's CT. This article presents (1) a design case that outlines specific challenges associated with the design, validation, and administration of early childhood CT assessments with special attention focused on fairness (i.e., equity) and (2) an explanation for how we dealt with these challenges in the design of an interview-based assessment of CT for kindergarten students.

In the sections that follow, we first lay out the scope of what is known about CT assessments, highlighting the critical issue of validity and describing the ECD approach to validity by design. Then, we describe our project, called Coding in Kindergarten, contextualizing our assessment design in the overall design process and aims of the project, including a summary of our procedures and sample. Then we turn to the central contribution of the paper and align the first three layers of ECD – domain analysis, domain modeling, and conceptual assessment framework – to our process of developing our Kindergarten CT assessment. Due to space limitations and our interest in illustrating our process and argument in sufficient depth, we focus on algorithmic thinking (AT), one sub-component of the larger CT construct. For AT, we describe (a) concepts and considerations that emerged from our domain analysis, (b) how these were articulated in terms of a design pattern, and (c) instantiated in the student, evidence, and task models in our assessment. Finally, we discuss how developing an early CT assessment using the iterative ECD process required us to engage with related important competency areas, such as spatial reasoning. We suggest some implications of this and other special considerations for designing assessments of kindergarten-level CT that bear on fairness and equity.

Current scope of assessing computational thinking

The computer science education community is growing rapidly and beginning to construct nuanced definitions of CT (e.g., Cutumisu et al., 2019; Grover et al., 2015; Shute et al., 2017; Tang et al., 2020) however, there lacks comprehensive research on how to measure CT. Part of the challenge is that CT has been conceptualized somewhat differently at different age ranges, and therefore different frameworks have been employed to both define and understand CT (e.g., Basu et al., 2020; Bers et al., 2014, 2019; Brennan & Resnick, 2012; Grover & Pea, 2013, 2018; Lye & Koh, 2014; Shute et al., 2017; Snow et al., 2019; Yadav et al., 2014). Partly due to these different conceptions and frameworks adopted to guide research on CT, current assessment tools for CT look very different from one another. Each has been created in a variety of formats, engage students in different tasks types, and have been developed to function in different educational settings. We aim to develop an assessment that can be used across different research projects studying the integration of CT into kindergarten classrooms. Specifically, to be used with young children who are preliterate or emerging readers.

Validity and reliability of computational thinking assessments

Assessment of CT has garnered enough interest and initial efforts from researchers, that we have begun to see systematic literature reviews on the topic (Cutumisu et al., 2019; Tang et al., 2020). After combing through well over 100 combined articles that empirically reported the use of assessments to measure CT, one of the most resounding shared findings was that the reliability and validity of the current CT assessments was in short order. For example, Tang et al. (2020) reported that of the 96 empirical CT assessment studies analyzed, only 45% reported reliability measures and only 18% reported validity evidence.

The role of validity in assessment design

Validity is a central issue in assessment design. Validity in assessments refers to the degree to which evidence and theory support the interpretations of test scores entailed by proposed uses of tests (American Educational Research Association, American Psychological Association & National Council on Measurement in Education [AERA, APA, NCME], 2014; Messick, 1989, 1995). Much like theory development, the process of validation involves accumulating relevant evidence to provide a sound scientific basis for the proposed interpretations of the assessment (American Educational Research Association, American Psychological Association & National Council on Measurement in Education [AERA, APA, NCME], 2014). However, establishing validity is an ongoing process and it may change as interpretations and uses develop or as new evidence is accumulated (American Educational Research Association, American Psychological Association & National Council on Measurement in Education [AERA, APA, NCME], 2014; Kane, 2006; M. T. Kane, 2013). It is important to stress that validity is not a property of the test but of the proposed interpretations and situated inferences of the scores (M. T. Kane, 2013). Thus, not only does validity ensure a test is measuring what it is intended to measure, it ensures inferences based on the outcome of a particular use of an assessment are appropriate.

Mislevy (2007) has argued for the need to make the underlying principles of assessment design more explicit and to structure such designs around assessment arguments. He referred to this as "validity by design" where "as test creators, we are not carrying out validation activities but carrying out design activities structured in such a way that validity evidence emerges" (Mislevy, 2007, p. 467).

Evidence-centered design & validity (by design)

Assessments require a comprehensive framework for making valid inferences about learning. One such framework is Evidence-Centered Design (ECD), which provides rigorous procedures for linking theories of learning and knowing to observations and to interpretation (Mislevy & Haertel, 2006). The ECD framework is a systematic way to design assessments and involves constructing educational assessments in terms of evidentiary arguments (Mislevy & Haertel, 2006). Evidentiary argumentation is borrowed from Toulmin's argument schema (Toulmin, 1958) where the argument is constructed through a series of logically connected propositions that are supported by data and subject to alternative explanations (Mislevy & Riconscente, 2005). The ECD framework provides the structure for developing the argument, but it is up to the assessment designer to provide the content (Mislevy & Riconscente, 2005). This view of assessment as argument is central to validity arguments presented by Kane (2006), M. T. Kane (2013)) and the AERA/NCME Standards (American Educational Research Association, American Psychological Association & National Council on Measurement in Education [AERA, APA, NCME], 2014).

ECD is a multilayer approach comprised five layers: (1) domain analysis, (2) domain modeling, (3), the conceptual assessment framework, (4) assessment implementation, and (5) assessment delivery. In layers 1 and 2, the focus is on the purposes of the assessment, the nature of knowing, and structures for observing and organizing knowledge. In the third layer, assessment designers focus on the student model (what skills are being assessed), the evidence model (how are skills measured), and the task model (situations that elicit the behaviors/evidence). These aspects of the design are interrelated. The ECD method was developed to assess complex performances. For this reason, other groups (e.g., Basu et al., 2020; Snow et al., 2019) have used ECD to develop assessments that measure CT practices. While Snow et al. (2019) documented some of the challenges of using ECD such as time and cost, they have also shown the potential of ECD as a framework for developing an assessment for CT practices. In particular, they have illustrated the importance of design patterns in that they provide a framework that clearly outlines the focal KSAs and guidance on how to develop tasks to measure them. Building off the work of Snow et al. (2019), we used ECD to guide the design of our assessment.

The coding in kindergarten project

The larger National Science Foundation-funded research (Grant no. NSF #1842116) from which this article originates, the Coding in Kindergarten (CiK) project, is investigating how to both integrate computer science into kindergarten mathematics instruction using commercial screen-free coding toys and assess the students' CT. We are using commercial toys for two reasons. First, a number of early childhood education classrooms, including those with which we partner, have a "no screen-time" policy, making apps that have been thoughtfully designed (e.g., Scratch Jr.) ineligible for kindergarten classroom use. Some work has been done to develop high quality academic research-based toys, but teachers and schools can view those as a substantial material investment beyond what they can typically afford (e.g., Bers et al., 2014). That has made easily purchased and easily replaced commercial toys an attractive option for educators as a way to introduce coding in a playful and simple way.

Second, and in line with the project's computer science and mathematics integration, we hypothesize these toys, which all involve moving a robot-like entity through a grid space, will foster spatial reasoning skills (Sarama & Clements, 2009). For example, prior research on young children's spatial assembly skills with blocks suggests an important link between spatial and mathematical skills and that tangible block-building activities facilitate the use of spatial language in cooperative social settings (Verdine et al., 2014). Similarly, these types of coding toy contexts engage students in cooperative spatial orientation and measurement experiences (Shumway et al., 2019; see Figure 1).

The CiK project has three goals: to operationalize what CT looks like in kindergarten classrooms when students interact with coding toys; to develop curricula and resources for teachers to use coding toys in their classrooms; and, to develop an assessment that could be used across toys that measures kindergarteners' CT. Thus, we are not assessing the curriculum per se, but the kinds of CT skills that are afforded by using screen-free coding toys and other similar learning objects (e.g., the Robot Turtles educational board game produced by ThinkFun). By necessity, we have had to design and implement curriculum as we operationalize CT and develop our assessment. Our research involves working in kindergarten classrooms in public schools. While we developed and piloted curriculum around the toys in classrooms (Silvis et al., 2020), the focus of this paper is on the design of our assessments and the validity evidence that has emerged in the early design process. Nonetheless, it is important to acknowledge the concurrent design activities, including the curricula and how enacting it in classrooms informed our definitions of CT, our design patterns, our student models, task models, and evidence models.

Figure 1. Group of Students and Teacher Interacting with Coding Robots.

Context of the research

We worked with four elementary schools in the Mountain West, United States. All four schools are classified as Title I; three are classified as rural, and those three also received funding to offer full-day kindergarten. Figure 2 presents the timeline of our work with the four schools, depicted as A, B, C, and D. In school A, we worked with three teachers and six classes. In schools B, C, and D we worked with one teacher and class per school.

Curriculum implementations involved a member of the research team, a former elementary or preschool teacher, teaching a 30-min lesson with a coding toy to small groups of 4–5 kindergarten students during their STEAM Centers (Science, Technology, Engineering, Art, and Mathematics activities that students visit in a rotation each week). Each lesson was observed and video recorded by a member of the research team. Assessment implementations involved a member of the research team working one-on-one with a child.

During *Phase 1*, Spring, Year 1, we piloted prototypes of curriculum tasks around two toys and then some assessment items. During *Phase 2*, we conducted two different types of pilots. Our first approach involved School A and was focused on piloting assessment items. The teachers, who had been using coding toys prior to working with us, had access to our curricula and resources and implemented coding lessons on their own. Members of the research team piloted assessment items with students. Our second approach focused on piloting both curriculum and assessments. In school B and C, members of the research team taught a sequence of six lessons around two or three different coding toys. At the end of the curriculum, researchers administered a version of our assessment with each student.

Data collection at School D was not completed at the time of writing this paper. For the purposes of this paper, we focus only on *Phase 2* in Figure 2, the highlighted period between early fall and late winter. We describe the *Phase 2* procedures and sample below. *Phase 3* (not pictured) involves scaling up curriculum implementation through professional development and continuing co-design of lessons using coding toys. While we expect teachers will deliver the lessons, the intended use of the assessment is as a tool for researchers who study early childhood CT. Although it will be available for teachers to use.

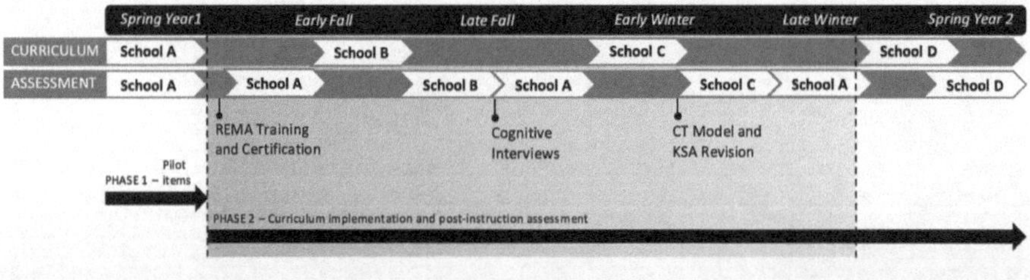

Figure 2. Timeline for Assessment Events Over a One-Year Period.

Sample for this study

Data in this paper focuses on six classrooms within three Title 1 schools (A, B, and C). It is important to note that full day kindergarten is rare in this region. The three schools we worked with implemented full day kindergarten in order to provide opportunities for students demonstrating academic risk upon kindergarten entry. Participants' parents were asked to complete an optional demographic survey, but not all of them did. Eighty-nine students took versions of the assessment during this time period (females = 45, males = 44). Forty-seven identified as White, 14 as Latinx, 1 as Black, 4 as Asian/Pacific Islander, 2 as Native American, and 2 as Other. Ten students reported receiving ESL services, 9 reported receiving special education services, and 22 as receiving free and reduced lunch.

Procedures for the assessment administration

Research team members were trained and certified on the administration of the Research-Based Early Mathematics Assessment, a validated mathematics assessment designed for a similar age group (Clements et al., 2019). The assessors all taught the curriculum in Schools B and C and demonstrated good rapport and management skills with children. They had prior formal classroom teaching experience and advanced degrees in education. On average, we were allotted 20 minutes to assess each child. Standard materials included in the assessment task environment were: a small 3-D printed moveable agent; a series of paper-based grids bound in a flip book that served as the navigation plane for moving the agent; a collection of paper-based arrow code tiles for sequencing and some preset laminated code strips; a program organizer to contain the sequence; administrator pages, with item scripts and prompts; and scoring sheets (see Figure 3). All assessments were video recorded and double scored.

Applying ECD to kindergarten CT assessment

As mentioned above, the ECD model has five layers; for the purposes of this paper, we discuss the first three in relation to our design process. Likewise, our assessment targets

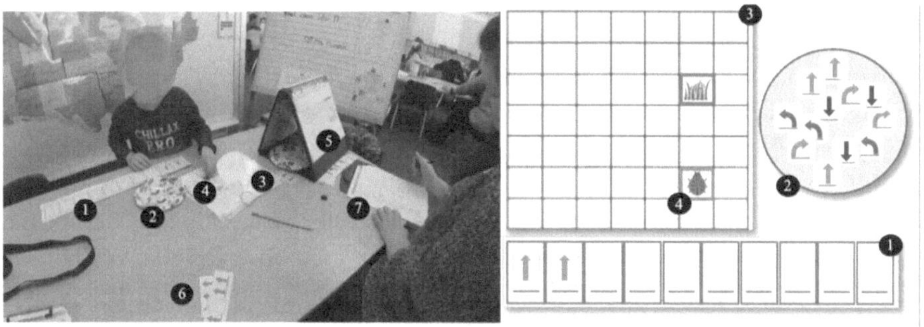

Figure 3. Student taking assessment. Left shows materials used in assessment tasks: (1) program organizer (2) arrow codes (3) grid pages, flip book (4) moveable agent (5) administration pages, with script (6) preset code strips (7) scoring sheets. Right shows child's-eye view of assessment materials.

multiple components of the CT construct, including algorithmic thinking, spatial thinking, debugging, and decomposition. Due to space limitations, we focus this paper on only the algorithmic thinking (AT) sub-component, and we document design changes that took place regarding this crucial component of CT over one year of an assessment design cycle. In what follows, we describe how our assessment and conceptualization of this construct developed and became further refined through ongoing domain analysis, domain modeling, and conceptual assessment re-framing.

Domain analysis

This layer of assessment design is focused on gathering information about the domain being assessed. At the core of this process is documenting "how this knowledge is acquired and used, as well as how competence is defined and how it develops" (Mislevy & Riconscente, 2005, p. 7). The information gathered is set up as arguments in the second layer, domain modeling. In this section, we summarize portions of our domain analysis: CT definitions and ways of measuring CT, AT, and critical aspects of AT assessment items (e.g., spatial thinking).

CT definitions

We started our domain analysis by looking at the published research on CT as well as assessments of CT (e.g., Angeli et al., 2016; Bienkowski et al., 2015; Brennan & Resnick, 2012; Shute et al., 2017; Snow et al., 2017, 2019; Sullivan & Bers, 2016; Weintrop et al., 2016). First, we culled information on definitions and frameworks for CT including the CSTA standards. We categorized the definitions based on context, concepts, and age.

Definitions of CT vary and are associated with the environments and domains in which particular research is situated. For example, studying CT the context of the Scratch programming environment, led Brennan & Resnick to identify concepts like sequences and practices like debugging as relevant for CT. Ehsan and Cardella (2017) framed CT within engineering education and identified CT competencies such as abstraction, algorithm and procedure, debugging/troubleshooting, pattern recognition, and simulation while observing first-grade students complete an engineering design task in an informal learning context. Weintrop et al. (2016) performed a review of CT literature and interviewed experts in mathematics and science in order to develop a taxonomic definition of CT consisting of four categories: (1) data practices (2) modeling and simulation practices (3) computational problem-solving practices and (4) systems thinking practices. When it comes to operationalizing CT- which factors? how many? how do they interact? what sorts of practices do they support?- context appears to matter a great deal.

In the domain of early childhood CS education, a number of researchers emphasize *coding and programming environments* as relevant for CT development, recognizing that other types of CT engagement beyond coding are possible (i.e. unplugged or story-based activities). Based in this emerging context for CT, we aligned our definition with the ways in which early childhood CT has been conceptualized and assessed, as a multi-factor construct (Bers, 2018; Relkin et al., 2020; Shute et al., 2017). Shute et al. (2017) developed a definition for CT as the "conceptual foundation required to solve problems effectively and efficiently" (p. 151). Included with their definition, they analyzed various CT concepts and categorized commonly identified CT skills into six main facets: decomposition, abstraction, algorithm design, debugging, iteration, and generalization. Similarly, Relkin et al. (2020) created the *TechCheck* which was specifically meant to assess young children's "various domains of CT described by Bers (2018) as developmentally appropriate for young children: algorithms, modularity, control structures, representation, hardware/ software, and debugging, with the exception of the design process" (p. 4). Though the operative dimensions may vary, a variety of CT assessments measure sub-constructs of CT and situate their measures in programming contexts.

Computational thinking assessments for young children: unique challenges

Currently, there is a lack of valid and reliable CT assessments for young children. One potential reason is because within the early-childhood literature, there has not been a consensus on a guiding framework, definition, or set of proxies with which to measure CT (e.g., Angeli & Valenides, 2019; Bers et al., 2019; Cittá et al., 2019; Martinez et al., 2015). For example, Angeli and Valenides (2019) measured CT of 50, five- to six-year-old children by individually assessing decomposition, AT, sequencing, and debugging and used these skills and practices to garner a holistic score of students' CT. Cittá et al. (2019) measured CT of 92, six- to ten-year-old students by having them take a paper-pencil test that assessed the " … students' ability to write and interpret an algorithm on paper in a closed environment represented by a chessboard" (p. 5). Bers et al. (2019) measured CT of 172, three- to five-year-old children by having them solve a robot-based challenge. The resulting product was assessed by the researcher using a Solve-Its checklist (Strawhacker et al., 2013). This checklist has progressive levels of complex programs, and as students include more or less complexity in their program, they receive more or less points. These three examples highlight how different CT assessments for young children measure different skills and practices of CT. The varying conceptions of what CT is and what skills and practices represent it, makes comparisons of CT learning challenging and demonstrates a need for a CT assessment aimed at generality. Based on converging literature, we defined CT as being comprised of the following facets: (a) algorithmic thinking, (b) decomposition, and (c) debugging. We added an additional skill (d) spatial reasoning, which is not part of most CT definitions but something we knew would be important based on the movement of the toys and the data collected thus far in the project.

A focus on the algorithmic thinking component of CT

There are different approaches and definitions to defining algorithms and programming in the literature. Most of them start out similarly by defining it as a step-by-step process to complete a task (e.g., CSTA CS standards, 1A-AP-08, 2017) and "design(ing) logical and ordered instructions for rendering a solution to a problem" (Shute et al., 2017, p. 12). These steps can be carried out with or without a computer. Algorithmic thinking (AT) can be thought of as the writing or reading of a sequence of steps to solve a problem, either computer-based or in an unplugged environment. Sometimes, this ability to read or write a program is measured directly (e.g., Cittá et al., 2019) and sometimes underlying skills such as sequencing (Bers et al., 2019; Martinez et al., 2015) or action-symbol correspondence (Muñoz-Repiso & Caballero-González, 2019) have been measured with young children. AT necessitates careful examination of items that would allow for a certain AT trajectory to be developed and assessed. This could include foundational skills like sequencing and action-symbol correspondence, but would not regard them as the sole indicators or algorithmic thinking. These foundational skills may be particularly critical to assess in young populations such as kindergarten.

In our review of standards and existing research, we initially defined algorithmic thinking in four parts: give a set of sequential instructions, correctly use the syntax of the coding system, create loops through loop procedures/control structures, and create functions. After three pilots of the AT assessment items (see Figure 2), we modified the aspects of assessment situations that can be varied to shift difficulty or focus, which we call variable features. After adjusting variable features in subsequent assessments, we were better able to observe what we thought kindergarten students could do.

As with the larger construct, the context in which we are operationalizing AT is relevant for measuring it. The context in which we are operationalizing and assessing AT is consistent with computational environments being widely adopted in early childhood settings: tangible coding materials that rely on a grid-based series of movement commands (e.g., Bers, 2018). While CT it is not exclusive to coding and programming, and not all coding and programming environments in early childhood involve grids and directional commands, this format is common enough- and the commercially available coding tools numerous enough- that we believe it is prudent to base early childhood CT assessment in this context. *However, circumscribing our assessment design around this particular context, introduced unexpected variables associated with a grid system and the development of spatial and directional thinking*. Rather than a limitation of our design or conceptualization of the constructs, we see this as a relevant finding for researchers and teachers who are trying to understand how CT operates and how to observe it in early childhood. We therefore analyzed the literature in this domain for its relevance to AT and CT.

Spatial thinking as a variable feature for AT items

Kindergarteners engage their spatial thinking to program toys and agents to move in a given space. Our curriculum tasks and AT assessment items were tethered to spatial thinking, and became an important variable feature of the AT assessment items. Hence, understanding student's knowledge of spatial relationships was particularly pertinent to our domain analysis. Spatial thinking entails understandings of space and objects' positions in space, reasoning with objects or representations in space, and operations on spatial relationships (National Research Council [NRC], 2006; Sarama & Clements, 2009). The K-12 mathematics curriculum relies on spatial thinking (National Research Council [NRC], 2006). The kindergarten geometry standards (Common Core State Standards Initiative [CCSSI], 2010) state, "describe objects in the environment using names of shapes, and describe the relative positions of these objects using terms such as above, below, beside, in front of, behind, and next to" (K.G.A.1). The standard emphasizes students' use of language to describe shapes and positions of shapes in space, but the critical underlying concepts are about developing students' spatial orientation. Spatial orientation is the understanding of different positions in space, and students first develop spatial orientation concepts in relation to their own position in space and later develop external-based reference systems using landmarks outside themselves (Sarama & Clements, 2009). In our programming context, students use their spatial orientation concepts when they move a toy or agent around a grid space. For example, in Figure 3, the student is holding the agent and ready to move the agent two movements forward on the grid to the grass.

Previous research suggests that mental rotations are difficult for young children and the ones who have more advanced mental rotation ability demonstrate higher levels of mathematical and computational thinking (Cheng & Mix, 2014; Cittá et al., 2019; Cuneo, 1985). Further, research also suggests young children progress through a specific spatial thinking learning trajectory (Sarama & Clements, 2009) and shift from certain egocentric to allocentric reference frames while programming (Clarke-Midura et al., 2021). In the Figure 3 example, the student is completing an easier task that involves two forward movements and is oriented in the same direction as the agent. The same task (two forward movements from the starting point to the grass) with the agent facing the student is more challenging and requires the student to consider the toys's frame of reference.

The grid-based system brings important challenges to our assessments, and it is important to consider the ways that students will be able to navigate movements within rows and columns on a grid. Our grids are two-dimensional arrays made up of rows and columns of squares. Student understanding of the arrangement of a grid, specifically the organized structure of the rows and columns of squares, requires the cognitive feat of spatial structuring. Spatial structuring is a form of abstraction that involves the ability to organize and coordinate a set of objects in space (Battista et al., 1998). In our context, kindergarteners are still developing spatial structuring concepts and understanding the organization of rows and columns in a grid is quite challenging until about the age of seven years old (Sarama & Clements, 2009). Hence, while the grid can provide an organizing structure to plan paths of movements for the toy or agent, young children may have difficulty understanding the grid's structure.

Domain modeling of AT

This layer of ECD is focused on diagramming or otherwise systematically planning how designers might create an assessment of the construct of interest. Modeling the domain means bringing the construct into alignment with the population, knowledge, behaviors, work products, and tasks (Oliveri et al., 2019). Along with operationalizing relevant attributes of the knowledge and population of interest that emerged from domain analysis, a major focus of articulating the domain model should be the evidence which supports the argument for the assessment's validity.

Design pattern for early childhood algorithmic thinking

As an approach to design modeling, design patterns (DPs) organize a set of decisions for developing assessments (Mislevy & Haertel, 2006). They consist of a series of defined attributes, such as focal knowledge, skills, and abilities (KSAs) and characteristic features of items, that structure an assessment. Design patterns rationalize an assessment's argument by describing how the construct of interest will be operationalized, elicited, and observed (Mislevy & Haertel, 2006). DPs also show how attributes of the design are interconnected and are a means of reconciling dependencies between nested design considerations (Oliveri et al., 2019). Domain modeling through the development of a design pattern took the form of the following attributes (see online supplemental material A for more detail on the DP).

Population and special considerations

In order to ensure fairness in assessments, designers must account for special considerations in a given target population (Oliveri et al., 2019). Examples of special considerations might be examinees' linguistic and cultural diversity, their prior knowledge, or other construct-relevant factors, like age or gender, that bear on making valid claims about test takers' performance. We were careful to create tasks that were not biased toward a particular socioeconomic status, culture, or prior background experiences. For example, we avoided holiday themes and activities that bias a particular group, and instead we used a bug theme among the physical materials and test items.

Our intended population is primarily preliterate, which demands certain considerations in assessment creation, such as interview-based delivery format, verbal instructions or observational assessment. Other issues that arise with young children is that they may or may not be familiar with typical symbols that seem customary to researchers, such as the rotate arrow (see Figure 3). In fact, these arrows may provoke different interpretations for children based on their previous, mundane experiences with the symbols (Silvis et al., 2020). For example, they may think the left rotate arrow means to make a rotate and forward movement, similar to what would be done in a car or walking around a corner, rather than make a stationary rotation.

We also subjected our measure to a time constraint of twenty minutes, knowing that there are limits on young children's attention and engagement that could influence their behaviors. In addition, screen-free and toy-free tasks were required, not only because teachers preferred these approaches, but also because we recognized that children have had variable exposure to programming contexts and different opportunities to learn to use coding toys. We considered children's' language development, both in terms of age (young children learning language) and linguistic background (English Language Learners), in the design of items and administration. Examples include using manipulatives (i.e., moving an agent on the grid and moving tangible arrows to a program organizer) and gestures (e.g., pointing to codes or a space on the grid), using prompts to support their responses, modifying questions for better comprehension, and eliciting nonverbal responses (e.g., yes/no questions, statements such as point to).

Assessment rationale

Building from definitions of CT that came out of our domain analysis, our initial working definition of AT contained four parts: give a set of sequential instructions; correctly use the syntax of the coding system; create loops through loop procedures/control structures; and create functions. Our original set of KSAs for AT were modeled after this definition. During early winter (see Figure 2), we found that this definition, while grounded in the literature on early CT, was inadequate for describing the kind of thinking kindergarteners exhibited when learning to code. For example, although our assessment is toy-free, the language of directional arrows is an inherent feature of most coding toys and programs geared towards children (Clarke-Midura et al., 2019). Thus, algorithms frequently take the form of an ordered sequence of directional arrow commands, the execution of which instructs an agent to navigate on a plane such as a map, grid, or number line.

Given the prevalence of this constraint in early childhood coding environments- and departing from broader conceptualizations of AT in the literature, we further refined the definition. For the purposes of our assessment, we define AT as developing and using logical and ordered sequences of instructions [see online supplemental material B]. By defining AT in this way, we retained sequencing at the center of the AT sub-component of the CT construct and ensured that assessment content would be context-relevant for the target population.

Focal knowledge, skills, and abilities (KSAs)

After piloting our assessment three times in the fall (two times with School A, one time with School B, see Figure 1), and after implementing the curriculum and observing what occurred under non-testing conditions, we modified our KSAs. Initially, the KSAs for AT were keyed to our four-part working definition, and the first two KSAs were that students could specify a short sequence of instructions (< 3) and a moderate length sequence of instructions (> 4). We subsequently unpacked and expanded our definition of AT into finer skills that were more reflective of the capabilities that students were demonstrating. One way we did this was to shift "length of sequence" from a variable skill to a variable task feature that could be modified to differentiate levels of AT. Another way was to more fully incorporate the requisite knowledge of one-to-one, movement-action correspondence into the KSAs for AT as a form of prior knowledge we were not (yet) measuring. We also articulated certain meta-computational concepts; test-taker knowledge of terms like "program", "code", and "instruction" had been implicit in student performance, but we now treated these as prior knowledge consequential for modeling and assessing AT.

Although our curriculum implementations and pilot assessments in late fall had begun to indicate that the ability to create functions and loops using control functions was beyond the skill level of kindergarteners we worked with, we preserved these as ceiling KSAs in our model and retained the corresponding tasks as "upper anchors" (Penuel et al., 2014, p. 79) in the assessment. We made this decision because we had only explored loops in classrooms with two of the toys to date. Admittedly, these two toys did not present loops and functions in a way that we found to be intuitive. However, Kibo, which we have not implemented in classrooms, presents loops as "repeat" blocks whose sequence begins with "Repeat begin" and ends with "repeat end". While research on Kibo has found that loops can be difficult for preschool-aged children (e.g., Elkin et al., 2018), we have not been able to conduct research on programming loops with Kibo in kindergarten classrooms. Further, Bers and colleagues, are exploring coding trajectories that they say start with sequences (order) and end with more complex patterns of sequences such as loops (Bers, 2019). Given that there is still much for the field to learn about coding in early childhood, we also plan to continue to explore how children engage with these concepts. Finally, in consideration of the fact that kindergarteners were in the process of learning to read and write, we realized that our working definition of AT foregrounded writing (specifying) code, but that reading and enacting programs were equally important. We therefore reorganized assessment items to align with these three activities in which students demonstrated the AT knowledge we were targeting. We provide a blueprint for how this was instantiated in our task design as part of the conceptual assessment framework (CAF).

Additional KSAs: the role of spatial reasoning

Another key insight from pilot assessments and curriculum implementations was that accounting for directional movements of an agent was central to warranting the assessment argument. In other words, alternative explanations of performance that hinged on students' directional and spatial knowledge emerged as a potential confound for our focal KSA. Teasing out the complex relationship between spatial and directional reasoning and disambiguating it from AT became a central design challenge.

One aim of assessment design patterning is to account for alternative explanations for performance. Design patterns are useful in this regard, because they provide a means of identifying additional KSAs implicated in the construct and determining whether these are collateral of, prerequisite for, or irrelevant to the construct (Oliveri et al., 2019). Whereas with older children, eliminating such sources of construct-irrelevant variance may have been an easier matter, young children's AT was entangled with their incomplete spatial knowledge of left and right, their varying interpretations of the semantics of arrow symbols, and their developing understanding of symbol-action correspondence.

We had to determine the relevance of spatial reasoning to the construct of interest. Was spatial reasoning an additional set of KSAs that needed explication in the design pattern? Or did we need to somehow incorporate it into our focal KSAs? Design patterns may identify additional "knowledge, skills, and abilities that are not part of the construct but depending on design choices and task features, may be required to perform the task" (Oliveri et al., 2019, p. 282). We decided to treat spatial reasoning in this way, as meaningfully relevant for solving the particular problems that arise in early childhood CT educational contexts. Our assessment includes a section of spatial reasoning items preceding the AT section, that, like the other components of CT has undergone revisions according to a series of design features. We discuss these in more detail following the kinds of work products and potential observations that we posited can provide evidence to support claims about student knowledge.

Potential work products and observations

Based on our definition of AT, we mapped KSAs onto student's potential work products and observable behaviors. We documented types of responses that could serve as indicators of their KSAs and evidence of latent proficiencies. For example, we expected that they could produce a program of three or more codes as well as programs with a turn. These programs, comprised of sequences of arrow cards arranged along a specially designed "program organizer," would serve as physical evidence of students' AT (see Figure 2).

Additional forms of evidence might be students' enactments of programs as they followed a pre-set program that we called a "code strip" (Figure 2). We knew from curriculum implementations and pilot assessments that enacting and simulating ready-made instructions was a primary mode through which students demonstrated their early knowledge of sequencing. We recognized that students learning to read English were likewise learning to read a directional arrow-based programming language and that to specify a sequence, they first needed to decode the symbol system. Enactments and think-aloud sessions with students provided means of accessing their ability to read and follow sequences. Furthermore, our focus on math integration suggested that counting the number of codes required to solve a problem served as another potential source of evidence of student's AT.

Variable features

We prototyped materials and piloted tasks that could elicit these potential work products and observable behaviors. Variable features are features of task sets that can be varied to change its level of difficulty, context (to facilitate new task generation), or granularity (Oliveri et al., 2019). Through iterative cycles of item redesign, piloting, and data analysis, we modified features of tasks to make them more or less difficult and to unpack AT. We identified variable features for the context of task sets: using a navigation plane such as a map, grid, or number line that the agent could move across; having a path that the agent takes with start and end points that correspond to the start and end of algorithms; manoeuvring an agent within the plane; positioning codes in a linear sequence; and embedding each task in a simple scenario representing a problem for the agent to solve.

Within this task set, we identified that three key program features, when varied, controlled the difficulty of tasks. The first was starting orientation of the program. When the agent shared a starting reference frame with the student, the task was easier (Shown later in Figure 5, Tasks 2 & 3). Conversely, a starting orientation that was rotated ninety degrees or was facing the child (rotated 180 degrees) increased the level of difficulty (Shown later in Figure 5, Task 1). The second program feature we varied was the presence of turns in the sequence. Tasks that asked children to generate or to follow a set of instructions were more challenging if they included turns in the sequence. Furthermore, increasing numbers of turns increased the difficulty level. The third program feature that we used to change difficulty level was the distance traveled; longer sequences presented more challenging problems.

The conceptual assessment framework

The third layer in the ECD approach is the conceptual assessment framework (CAF). This layer is a blueprint of the assessment where decisions are made in regards to the statistical models, materials and delivery processes that will inform students' work products, and the scoring (Mislevy & Riconscente, 2005).

The CAF is comprised of three models: the student model, the evidence model, and the task model. These three models are fleshed out with the information in the design pattern and work together to provide technical details of the assessment. The CAF specifies how items can be varied to create families of items and how we update claims about students' proficiencies based on their work products or performances. The CAF also "serves as another place for examining the impact the assessment may have on test takers from different populations" and to "minimize inadvertent construct-irrelevant demands" (Oliveri et al., 2019, p. 291)

Student model: what are we measuring?

The Student Model answers the question of what the assessment is measuring. It is the latent proficiency variables in the statistical model used to accumulate evidence about the student. The Student Model variables are the link between students' performances on tasks and the claims we wish to make about their proficiencies (Mislevy & Riconscente, 2005). In the present study, the student model for AT is a simple model with one student variable. In the model, we will accumulate a number right or total score to characterize students' overall proficiency in AT.

The evidence model: how do we measure it?

The Evidence Model provides the technical details on how the assessment measures the Student Model. In the Evidence Model we update the Student Model based on observations of what children say or do. The Evidence Model has two components: evaluation and measurement. The evaluation component is linked to the Student Model and puts a value on (evaluates) the observations of what students say and do. This can be done in the form of a rubric, answer key, etc. The measurement component compiles the data from evidence component across the tasks. While "each piece of data directly characterizes some aspect of a particular performance, it also conveys some information about the targeted claim regarding what the student knows or can do' (Mislevy & Riconscente, 2005, p. 19). In the present study, observations of students' actions on each task (item) get scored as correct or incorrect based on our answer key.

The task model: in what situation(s) do we measure it?

The Task Model describes the kinds of situations that will elicit the observations of students' proficiency. The Task Model is linked to the work products, characteristic features, and variable features of the tasks described in the design patterns. The claims we want to make about students guide and shape the design decisions around the task features. Such decisions include the form of the work product, what materials are necessary and directives. Many important decisions are made at this layer that affect the inferences that will be made from the test. For example, it is important there are: a sufficient number of tasks to provide information about the construct, sufficient opportunities for students to show the broad range of KSAs necessary to assess the targeted construct, and opportunities for students to demonstrate the focal KSAs in ways that suit the intended context of use (Oliveri et al., 2019).

These three models are connected and linked in an assembly model. The assembly model specifies how the Student Model, Evidence Model, and Task Model work together to generate sufficient evidence to form a valid assessment (Mislevy & Riconscente, 2005). Our plan to is use Item Response Theory (IRT), in particular, a Rasch model to fit the data. The Rasch model analyses will allow us to ensure the unidimensionality of AT and estimate the relative location of items and persons on a single scale (De Ayala, 2013). To date, we have not piloted our assessment with a large enough sample to conduct IRT on our AT data. However, we collected valid evidence on the internal structure of AT with 59 students in School A in late winter. Cohen's kappa was computed to test the agreement between two coders (i.e, the inter-rater reliability). Results showed that there was an almost perfect agreement between the two coders, Cohen's kappa = .99, $p < .001$. We conducted the Kuder-Richardson Formula 20 (KR20), an index of the internal consistency reliability for binary measurements, on our pilot measure of AT that contained 11 items. Our KR20 value was .82, which is in the acceptable range. Item analysis for this sample indicated a range of item difficulty from .13 to .83. Our goal is to include a range of tasks of varying complexity to discriminate students with different ability levels.

Developing the task models: three examples

Figure 4 presents our Student Model and example observations and task features. Below we present three different task models for our claim that students can specify a program using a sequence of codes [Figure 5]. In all of our items, we provide students with a grid that includes a green square indicating starting position and a red square indicating ending position (the green arrow indicates starting orientation for the assessor, but is not depicted on student grids). All items in our assessment involve an assessor interacting

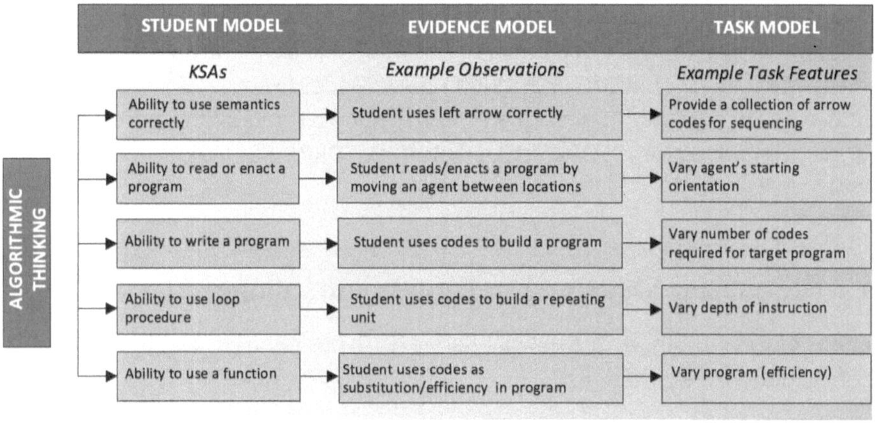

Figure 4. Student, Evidence, and Task Models for AT.

GRID IMAGE	TASK FEATURES AND MATERIALS	
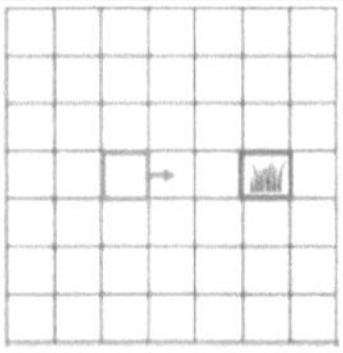	***Task Model Example 1:***	Specifying a sequence of code that gets agent from green square to red square.
	Variable Feature	Starting orientation is to the right.
	Administration	Orient agent, point to start and end goals
	Materials	Grid, Agent, Arrows
	Answer Key	**FORWARD, FORWARD, FORWARD**. Alternative paths accepted if it gets them from green to red.
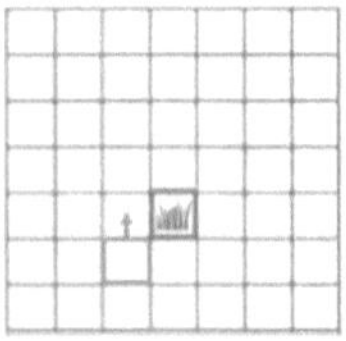	***Task Model Example 2:***	Specifying a sequence of code that gets agent from green square to red square.
	Variable Feature	Sequence involves a turn
	Administration	Orient agent, point to start and end goals
	Materials	Grid, Agent, Arrows
	Answer Key	**FORWARD, ROTATE RIGHT, FORWARD**. Alternative paths accepted if it gets them from green to red.
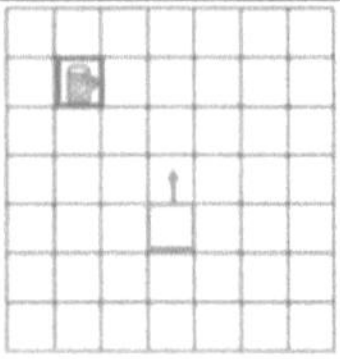	***Task Model Example 3:***	Specifying a sequence of code that gets agent from green square to red square.
	Variable Feature	Distance traveled > 3; sequence involves a turn
	Administration	Orient agent, point to start and end goals
	Materials	Grid, Agent, Arrows
	Answer Key	**FORWARD, FORWARD, FORWARD, ROTATE LEFT, FORWARD, FORWARD.** Alternative paths accepted if it gets them from green to red.

Figure 5. Task Model Examples.

with a student by reading a question and emphasizing materials or gesturing to the grid. In the following three examples, we accept alternative correct sequences; however, for the purposes of this paper, we provide only the most common (most efficient) response option. Please see online supplemental material D for an Assessment Administration example.

Task Model Example 1 varies the starting orientation for a short distance traveled. The task features for this model involve a starting orientation where the agent is facing towards the right from a students' perspective. For this particular task, the most efficient correct answer is three forward arrows (i.e. FORWARD, FORWARD, FORWARD). Task Model Example 2 varies the presence of a turn. For this particular item, the answer we are looking for is FORWARD, ROTATE RIGHT, FORWARD. Task Model Example 3 varies two features: traveling a distance greater than 3 squares and making a turn. For this particular item, the answer we are looking for is FORWARD, FORWARD, FORWARD, ROTATE LEFT, FORWARD, FORWARD.

Evidence to support arguments of fairness and validity

At the time of this writing, we are in what Kane (2010) has referred to as the *developmental stage* of validation. He writes:

> as an assessment program is being developed, the developers are expected to produce materials and procedures that support the proposed interpretations and uses and to make a case for the validity of the proposed interpretations and uses, and it is appropriate to talk about their efforts 'to validate' the claims being made. (Kane, 2010, p. 4)

While accumulating the relevant evidence to provide a sound scientific basis for the proposed interpretations of our assessment, we constantly ask *what is being claimed* and, given the evidence we have accumulated, *are the claims warranted for all students*? Using ECD provided some structure for us to map out evidence of how students' performances support the inferences that are made regarding students CT competency.

During the task design process, we made several decisions focused on assessing and identifying fairness (Oliveri et al., 2019). For example, we carefully selected the scenarios and vocabulary so that the context and content is accessible to all kindergarten students. We tried to provide opportunities for students to demonstrate the focal KSAs in ways that suit the intended context of use (i.e., kindergarten classrooms). We also considered possible alternative explanations that might account for the observed performances and scores on the assessment.

We collected validity evidence based on student responses through pilots and cognitive interviews with kindergarten students in public school classrooms, the population for which our assessment is intended. The cognitive interviews allowed us to identify any hidden assumptions or alternative plausible interpretations of the test scores. These data also allowed us to design counter arguments to our claims that the assessment is measuring what it is intended to measure. We gathered evidence of construct underrepresentation and expanded our definition of CT to account for spatial reasoning. We expanded our definitions of AT to account for the reading, writing, and enacting of programs. We also built counter arguments around construct irrelevant variance. In doing so, we modified our vocabulary and the language used in our tasks.

These efforts are guiding us towards our larger evaluation of our assessment where we will move to what Kane calls the *appraisal stage* (Kane, 2010). In this stage, we will calibrate our assessment and gather sufficient data to conduct a "critical evaluation of the extent to which the proposed interpretations and uses are plausible and appropriate" (Kane, 2010, p. 4).

Discussion

A central contention motivating this article is that there is a growing need to develop valid assessments of CT for kindergarten. At the same time, it is a challenging undertaking because so much has yet to be specified, both in terms of the types of instruction that are emerging and in the constructs that we hope to eventually measure. The goal of this paper has been to present a case of our efforts to use ECD to develop, test, and refine assessment items for the construct of AT with kindergartners. Design cases are useful for others interested in assessment as it can provide some procedural guidance for how to use ECD for the design of assessments in computer science education. At the same time, cases are useful for surfacing domain-specific tensions and challenges. It is to three of those challenges we now turn our discussion.

Challenge 1: assessing a nebulous construct

Typical assessment design processes presume that some image of how a target construct is instantiated can serve as a blueprint for the assessment development process. However, that image is lacking for early childhood CT. Because research does not yet converge on precisely what to assess and how to assess CT in early childhood, we started our project by operationalizing CT as best as we could give existing literature and standards documents. Through the iterative process of designing and piloting assessment items with kindergartners who have participated in some introductory toy-based CT instruction, we have necessarily adjusted both our items and our understanding of the underlying AT construct. The KSAs we articulated changed such that our initial targets became upward boundaries (e.g., looping) and more specific KSAs that were developmentally appropriate were articulated and added (e.g., spatial reasoning). Moreover, we expanded our KSAs to be inclusive of reading and enacting existing algorithms rather than focus exclusively on the production of algorithms. Because we iterated, we were able to begin to produce what we believe to be a more accurate image of what kindergartner's AT looks like when instantiated in these sorts of grid-based navigation tasks.

Challenge 2: expanding our definitions of CT

Many of the newly emerging tools and educational toys that are being used for kindergarten CT instruction involve navigating an agent through two-dimensional grid space (Clarke-Midura et al., 2019). Taking into account the current lack of consensus in the field as to what fully constitutes AT, and by extension, CT, it has still been an important realization for us that spatial reasoning plays a major role in young children's CT. Thus, for our assessment development process, we have found it appropriate to expand the scope of CT to include spatial reasoning. For kindergarteners, this involves both working with different spatial orientations – for instance, understanding right and left when an agent is facing the same direction as the child and when it is facing a different one – and spatial reasoning – such as when a student is trying to move an agent through the 2D grid space and needs to differentiate between rotations and translations. We came to this realization in our early pilot tests of our CT assessment items and began seeing systematic errors that students were making. By creating and testing new, related items, we have been able to confirm that children's spatial thinking plays an important role in these kinds of common CT tasks.

Challenge 3: assessing preliterate children

Given that the target population we are trying to assess is preliterate, our assessment protocol needed to be in the form of one-to-one oral administration. This puts it in the same style as other established assessments for early childhood (e.g., REMA, Clements et al., 2019) in terms of use of manipulatives, interview assessment, and prompts to aid task comprehension. There are advantages and drawbacks of this approach, with implications for scaling up administration. Although the assessment is intended as a research tool, teachers who may want to administer items would require brief training and time for one-on-one interviewing of their students (similar to how most assessments administered in kindergarten classrooms). We also recognize that children's developing oral and verbal abilities may influence how they interpret and respond to items, and we have taken steps such as cognitive interviewing a small subset of children to maximize clarity of scripts and prompts. Despite these constraints, we believe interview style assessment is most appropriate for Kindergarten children, because they are preliterate.

While we tried to make the assessment items accessible to young children, we did discover that their different levels of syntactic and semantic knowledge could affect performance as well. Arrows, which are commonly used as individual code instructions for this age group, are not necessarily transparent as a representational convention. Moreover, the importance of sequencing codes in a consistent manner, such as individually from left to right, and with a one-to-one mapping of a code onto a single increment of movement, cannot be taken for granted. To better understand students' AT, we revised and included items that ask children to read and enact existing programs rather than only write a program (specify a sequence of codes). By using items that provide a window into students' AT in terms of reading, enacting, and specifying a sequence of codes, we were better able to see what kindergarten students were able to do and which skills they were still developing.

Conclusion

Assessment design is influenced by its purpose, the context in which it will be used, and practical constraints such as resources and time (Pellegrino et al., 2016). In this paper, we presented how we are using ECD to develop an assessment of kindergarten-aged children's CT. In particular, we focused on our early design processes and decisions around operationalizing AT for kindergarten-aged students. We shared some of the realizations that we have made about AT for this population, such as the role that spatial reasoning plays in how AT is currently instantiated with newly developed coding toys and learning materials. While we are designing our assessment to be usable in conjunction with a broad set of instructional resources, we believe that our operationalizations are useful for others who may choose to develop their own assessment instruments for comparable populations. Up until now, there has not been a consensus on a guiding framework, definition, or set of proxies in which to measure CT in early childhood (e.g., Angeli &

Valenides, 2019; Bers et al., 2019; Cittá et al., 2019; Martinez et al., 2015; Snow et al., 2019) nor has there been consistent inclusion of spatial reasoning in definitions of CT.

The process we have described for AT item development is comparable to what we have been undertaking with other facets of CT, such as debugging and decomposition (see online supplemental material C). Our next steps are to finish refinements on our instrument, perform trials with more students, and apply IRT in order so to generate evidence that the assessment can effectively fill the niche it is designed to fill. It is our hope that as this work progresses, and other groups also pursue important work with this age group, that we will soon be in a stronger position for both characterizing and measuring the CT of young children.

Disclosure statement

No potential conflict of interest was reported by the authors.

Funding

This work was supported by the National Science Foundation grant #1842116. Any opinions, findings, and conclusions or recommendations expressed in this paper are those of the authors and do not necessarily reflect the views of the National Science Foundation or Utah State University.

ORCID

Jody Clarke-Midura http://orcid.org/0000-0001-5434-0324
Deborah Silvis http://orcid.org/0000-0001-5139-9048
Jessica F. Shumway http://orcid.org/0000-0001-7655-565X

Victor R. Lee http://orcid.org/0000-0001-6434-7589
Joseph S. Kozlowski http://orcid.org/0000-0003-4163-2955

References

American Educational Research Association, American Psychological Association, & National Council on Measurement in Education [AERA, APA, NCME] (2014). *Standards for educational and psychological testing*. APA

Angeli, C., & Valanides, N. (2019). Developing young children's computational thinking with educational robotics: An interaction effect between gender and scaffolding strategy. *Computers in Human Behavior, 105*, 1–13. https://doi.org/10.1016/j.chb.2019.03.018

Angeli, C., Voogt, J., Fluck, A., Webb, M., Cox, M., Malyn-Smith, J., & Zagami, J. (2016). A K-6 computational thinking curriculum framework: Implications for teacher knowledge. *Journal of Educational Technology & Society, 19*(3), 47–57. https://www.jstor.org/stable/jeductechsoci.19.3.47

Basu, S., Rutstein, D., Xu, Y., & Shear, L. (2020, February). A principled approach to designing a computational thinking practices assessment for early grades. In *Proceedings of the 51st ACM Technical Symposium on Computer Science Education* (912–918).

Battista, M. T., Clements, D. H., Arnoff, J., Battista, K., & Van Auken Borrow, C. (1998). Students' spatial structuring of 2D arrays of squares. *Journal for Research in Mathematics Education, 29*(5), 503–532. https://doi.org/10.5951/jresematheduc.29.5.0503

Bers, M. U. (2018). *Coding as a playground: Programming and computational thinking in the early childhood classroom*. Routledge. https://doi.org/10.4324/9781315398945

Bers, M. U. (2019). Coding as another language: A pedagogical approach for teaching computer science in early childhood. *Journal of Computers in Education*, *6*(4), 499–528. https://doi.org/10.1007/s40692-019-00147-3

Bers, M. U., Flannery, L., Kazakoff, E. R., & Sullivan, A. (2014). Computational thinking and tinkering: Exploration of an early childhood robotics curriculum. *Computers & Education*, *72*, 145–157. https://doi.org/10.1016/j.compedu.2013.10.020

Bers, M. U., González-González, C., & Armas-Torres, M. B. (2019). Coding as a playground: Promoting positive learning experiences in childhood classrooms. *Computers & Education*, *138*, 130–145. https://doi.org/10.1016/j.compedu.2019.04.013

Bienkowski, M., Snow, E., Rutstein, D. W., & Grover, S. (2015). *Assessment design patterns for computational thinking practices in secondary computer science: A first look (SRI technical report)*. SRI International. http://pact.sri.com/resources.html

Brennan, K., & Resnick, M. (2012). Using artifact-based interviews to study the development of computational thinking in interactive media design. *Paper presented at annual American Educational Research Association meeting*, Vancouver, BC, Canada.

Brennan, K., & Resnick, M. (2012, April). *New frameworks for studying and assessing the development of computational thinking*. In Proceedings of the 2012 annual meeting of the American educational research association (Vol. 1, p. 25). Vancouver, Canada.

Cheng, Y. L., & Mix, K. S. (2014). Spatial training improves children's mathematics ability. *Journal of Cognition and Development*, *15*(1), 2–11. https://doi.org/10.1080/15248372.2012.725186.

Cittá, G., Gentile, M., Allegra, M., Arrigo, M., Contri, D., Ottaviano, S., Reale, F., & Sciortino, M. (2019, November). The effects of mental rotation on computational thinking. *Computers & Education*, *141* Article number 103613. https://doi.org/10.1016/j.compedu.2019.103613

Clarke-Midura, J., Kozlowski, J., Shumway, J.F., and Lee, V.R. (2021). How young children engage in and shift between reference frames when playing with coding toys. *InternationalJournal of Child-Computer Interaction*. https://doi.org/10.1016/j.ijcci.2021.100250

Clarke-Midura, J., Lee, V. R., Shumway, J. F., & Hamilton, M. M. (2019). The building blocks of coding: A comparison of early childhood coding toys. *Information and Learning Sciences*, *120*(7/8), 505–518. https://doi.org/10.1108/ILS-06-2019-0059

Clements, D. H., Sarama, J., Wolfe, C. B., & Day-Hess, C. A. (2019). *REMA—research-based early mathematics assessment*. Kennedy Institute, University of Denver.

Common Core State Standards Initiative [CCSSI]. (2010). *Common core state standards for mathematics*. http://www.corestandards.org/Math/

Computers Science Teachers Association. (2017) . *K-12 computer science standards*. Creative Commons Attributions.

Cuneo, D. O. (1985). Young children and turtle graphics programming: Understanding turtle commands. *Paper presented at the Biennial Meeting of the Society for Research in Child Development*. Toronto, Ontario, Canada, April 25.

Cutumisu, M., Adams, C., & Lu, C. (2019). A scoping review of empirical research on recent computational thinking assessments. *Journal of Science Education and Technology*, *28*(6), 651–676. https://doi.org/10.1007/s10956-019-09799-3

De Ayala, R. J. (2013). *The theory and practice of item response theory*. Guilford Publications.

Ehsan, H., & Cardella, M. E. (2017). Capturing the computational thinking of families with young children. *Paper presented at the 124th ASEE Annual Conference & Exposition*. Columbus, OH.

Elkin, M., Sullivan, A., & Bers, M. U. (2018). Books, butterflies, and 'bots: Integrating engineering and robotics into early childhood curricula. In *Early engineering learning* (pp. 225–248). Springer, Singapore.

Grover, S., & Pea, R. (2018). Computational thinking: A competency whose time has come. In S. Sentence, E. Barendsen, & C. Schulte (Eds.), *Computer science education: Perspective on teaching and learning in school* (pp. 19–38). Bloomsbury Academic.

Grover, S., & Pea, R. (2013). Computational thinking in K-12: A review of the state of the field. *Educational Researcher*, *42*(1), 38–43. https://doi.org/10.3102/0013189X12463051

Grover, S., Pea, R., & Cooper, S. (2015). Designing for deeper learning in a blended computer science course for middle school students. *Computer Science Education*, *25*(2), 199–237. https://doi.org/10.1080/08993408.2015.1033142

Hamilton, M., Clarke-Midura, J., Shumway, J. F., & Lee, V. (2020). An emerging technology report on computational toys in early childhood. *Technology, Knowledge and Learning*, *25*(1), 213–224. https://doi.org/10.1007/s10758-019-09423-8

Kane, M. (2006). Validation. In R. Brennan (Ed.), *Educational measurement* (4th ed., pp. 17–64). Greenwood Publishing.

Kane, M. (2010). Validity and fairness. *Language Testing*, *27*(2), 177–182. https://doi.org/10.1177/0265532209349467

Kane, M. T. (2013). Validating the interpretations and uses of test scores. *Journal of Educational Measurement*, *50*(1), 1–73. https://doi.org/10.1111/jedm.12000

Lye, S. Y., & Koh, J. H. L. (2014). Review on teaching and learning of computational thinking through programming: What is next for K-12. *Computers in Human Behavior*, *41*, 51–61. https://doi.org/10.1016/j.chb.2014.09.012

Martinez, C., Gomez, M. J., & Benotti, L. (2015). A comparison of preschool and elementary school children learning computer science concepts through a multilanguage robot programming platform. *In Proceedings of the 2015 ACM Conference on Innovation and Technology in Computer Science Education,* (159–164), Vilnius, Lithuania.

Messick, S. (1989). Validity. In R. L. Linn (Ed.), *Educational measurement* (3rd ed., pp. 13–104). American Council on education and Macmillan.

Messick, S. (1995). Standards of validity and the validity of standards in performance asessment. *Educational Measurement: Issues and Practice*, *14*(4), 5–8.

Mislevy, R. J. (2007). Validity by design. *Educational Researcher*, *36*(8), 463–469. https://doi.org/10.3102/0013189X07311660

Mislevy, R. J., & Haertel, G. D. (2006). Implications of evidence-centered design for educational testing. *Educational Measurement: Issues and Practice*, *25*(4), 6–20. https://doi.org/10.1111/j.1745-3992.2006.00075.x

Mislevy, R. J., & Riconscente, M. M. (2005). *Evidence-centered assessment design: Layers, structures, and terminology* (PADI Technical Report 9). SRI International. https://padi.sri.com/downloads/TR9_ECD.pdf

Muñoz-Repiso, A. G. V., & Caballero-González, Y.-A. (2019). Robotics to develop computational thinking in early childhood education. *Comunicar, 27*(59), 63–72. https://doi.org/10.3916/C59-2019-06

National Research Council [NRC]. (2006) . *Learning to think spatially: GIS as a support system in the K–12 curriculum*. The National Academies Press.

Oliveri, M. E., Lawless, R., & Mislevy, R. S. (2019). Using evidence-centered design to support the development of culturally and linguistically sensitive collaborative problem-solving assessments. *International Journal of Testing, 19*(3), 270–300. https://doi.org/10.1080/15305058.2018.1543308

Papert, S. (1980). *Mindstorms: Children, computers, and powerful ideas*. Basic Books Inc.

Pellegrino, J. W., DiBello, L. V., & Goldman, S. R. (2016). A framework for conceptualizing and evaluating the validity of instructionally relevant assessments. *Educational Psychologist, 51*(1), 59–81. https://doi.org/10.1080/00461520.2016.1145550

Penuel, W. R., Confrey, J., Maloney, A., & Rupp, A. A. (2014). Design decisions in developing learning trajectories-based assessments in mathematics: A case study. *Journal of the Learning Sciences, 23* (1), 47–95. https://doi.org/10.1080/10508406.2013.866118

Relkin, R., Ruiter, L. D., & Bers, M. U. (2020). *TechCheck*: Development and validation of an unplugged assessment of computational thinking in early childhood education. *Journal of Science Education and Technology, 29*, 482–498. https://doi.org/10.1007/s10856-020-09831-x.

Sarama, J., & Clements, D. H. (2009). *Early childhood mathematics education research: Learning trajectories for young children*. Routledge.

Shumway, J. F., Clarke-Midura, J., Lee, V. R., Hamilton, M. M., & Baczuk, C. (2019). Coding toys in kindergarten. *Teaching Children Mathematics, 25*(5), 314–317. https://doi.org/10.5951/teacchilmath.25.5.0314

Shute, V. J., Sun, C., & Asbell-Clarke, J. (2017). Demystifying computational thinking. *Educational Research Review, 22*, 142–158. https://doi.org/10.1016/j.edurev.2017.09.003

Silvis, D., Lee, V., Clarke-Midura, J., Shumway, J., & Kozlowski, J. (2020). Blending everyday movement and representational infrastructure: An interaction analysis of kindergarteners coding robot routes. In M. Gresalfi & L. Horn (Edited by.), *Proceedings of International Conference of the Learning Sciences (ICLS)* 2020 Nashville, TN: International Society of the Learning Sciences.

Snow, E., Rutstein, D., Basu, S., Bienkowski, M., & Everson, H. T. (2019). Leveraging evidence-centered design to develop assessments of computational thinking practices. *International Journal of Testing, 19*(2), 103–127. https://doi.org/10.1080/15305058.2018.1543311

Snow, E., Tate, C., Rutstein, D., & Bienkowski, M. (2017). *Assessment design patterns for computational thinking practices in exploring computer science*. SRI International.

Strawhacker, A., Sullivan, A., & Bers, N. U. (2013, June). TUI, GUI, HUI: is a bimodal interface truly worth the sum of its parts? In Proceedings of the 12th International Conference on Interaction Design and Children (pp. 309–312), New York, NY.

Sullivan, A., & Bers, M. U. (2016). Robotics in the early childhood classroom: Learning outcomes from an 8-week robotics curriculum in pre-kindergarten through second grade. *International Journal of Technology and Design Education, 27*(1), 3–20. https://doi.org/10.1007/s10798-015-9304-5

Tang, X., Yin, Y., Lin, Q., Hadad, R., & Zhai, X. (2020). Assessing computational thinking: A systematic review of empirical studies. *Computers & Education 148,* 103798. https://doi.org/10.1016/j.compedu.2019.103798

Toulmin, S. (1958). *The uses of argument*. Cambridge University Press.

Verdine, B. N., Golinkoff, R. M., Hirsh-Pasek, K., Newcombe, N. S., Filipowicz, A. T., & Chang, A. (2014). Deconstructing building blocks: Preschoolers' spatial assembly performance relates to early mathematical skills. *Child Development, 85*(3), 1062–1076. https://doi.org/10.1111/cdev.12165

Weintrop, D., Beheshti, E., Horn, M., Orton, K., Jona, J., Trouille, L., & Wilensky, U. (2016). Defining computational thinking for mathematics and science classrooms. *Journal of Science Education and Technology, 25*(1), 127–147. https://doi.org/10.1007/s10956-015-9581-5

Yadav, A., Mayfield, C., Zhou, N., Hambrusch, S., & Korb, J. T. (2014). Computational thinking in elementary and secondary teacher education. *ACM Transactions on Computing Education, 14*(1), 5. https://doi.org/10.1145/2576872

Design and validation of learning trajectory-based assessments for computational thinking in upper elementary grades

Brian D. Gane, Maya Israel, Noor Elagha, Wei Yan, Feiya Luo
and James W. Pellegrino

ABSTRACT
Background & Context: We describe the rationale, design, and initial validation of computational thinking (CT) assessments to pair with curricular lessons that integrate fractions and CT.
Objective: We used cognitive models of CT (learning trajectories; LTs) to design assessments and obtained evidence to support a validity argument
Method: We used the LTs and evidence-centered design to develop assessments that 144 Grade 3 and Grade 4 elementary students completed following the integrated instruction. We analyzed data using multiple psychometric approaches.
Findings: The design approach and data analysis suggest that the assessments are well-suited to evaluate students' CT knowledge, skills and abilities across multiple LTs.
Implications: We show how to use LTs to design assessments that can yield valid inferences about students' CT competencies; these methods can be adopted and extended by others to create additional assessments that can advance computer science education.

Introduction and background

Computer science (CS) education has rapidly expanded into the elementary grades. The rationale for this expansion often includes assertions that exposing elementary students to CS and computational thinking (CT) helps to develop their problem-solving skills (e.g., Voogt et al., 2015; Voskoglou & Buckley, 2012) and persistence during challenging tasks (e.g., Barr et al., 2011; Barr & Stephenson, 2011). This literature further suggests that engaging in these early and sustained computational experiences provides students an advantage given the ubiquity of computational technologies. Significant efforts are currently underway to develop a wide range of methods of assessing young learners' CT knowledge and skills in response to this expansion. These efforts include developing and using assessments that utilize automated tools such as Dr. Scratch (e.g., Moreno-León et al., 2015), Parson's puzzles (e.g., Denny et al.,

2008; Ericson et al., 2017), rubrics for evaluating computational artifacts (e.g., Seiter & Foreman, 2013), artifact-based interviews (e.g., Brennan & Resnick, 2012), and paper-and-pencil assessments of CT concepts (e.g., Román-González et al., 2017). While many CT assessments have been designed, only a small number of studies provided validity evidence (Tang et al., 2020), making it difficult for researchers and teachers to interpret existing CT assessments with confidence. In addition, most assessments lack a theoretical framework that consistently identifies which CT constructs are assessed and at what levels of sophistication. Therefore, it is crucial to use an assessment design approach that is theoretically grounded and guided by well-established design frameworks in order to fully unravel the complexities of students' CT skill development and to capture students' CT understanding (Tang et al., 2020). In this paper, we describe a CT assessment development effort that is theoretically grounded in three literatures: (1) CT literature, through the use of CT learning trajectories to specify the model of student knowledge, (2) assessment design literature, through the use of an evidence-centered design logic, and (3) assessment validation literature, using an argument-based approach for specification, analysis, and interpretation of relevant data.

Given that elementary CT education is still new, efforts to develop and implement CT assessments are taking place at the same time that researchers and practitioners are debating the definitions of CT, experimenting with creating developmentally appropriate instructional materials, and exploring ways to integrate CT into other content areas (e.g., mathematics and science). These complexities must be acknowledged in any discussion of developing, validating, and implementing CT assessments as they have significant implications for their design and use. In fact, development of CT assessments can advance these efforts because measurement of hypothetical constructs like CT provides an operational definition of the construct Crocker, 2006, as necessary for researchers as it is for practitioners.

Definitions of CT

CT is often defined as a mental process for formulating and solving problems in a manner that draws on concepts that are fundamental to computer science (Wing, 2006). This definition has been expanded and revised numerous times (Barr et al., 2011), with no single definition emerging to guide the development of CT instruction and assessment. In all likelihood, there is no single definition of CT that applies across all content areas and grade levels. However, Shute and colleagues (2017) synthesized CT definitions across multiple studies and frameworks and came up with a consolidated definition of CT, "The conceptual foundation required to solve problems effectively and efficiently (i.e., algorithmically, with or without the assistance of computers) with solutions that are reusable in different contexts" (p. 151). Although outside the scope of this paper, interested readers can refer to Shute et al. (2017) for an in-depth analysis of the competing definitions of CT.

The K-12 Computer Science Framework (2016) and the revised Computer Science Teachers Association [CSTA] (2017) define five computational concepts (i.e., computing systems, networks and the internet, data and analysis, algorithms and programming, and impacts of computing) and seven practices (fostering an inclusive computing culture, collaborating around computing, recognizing and defining computational problems, developing and using abstractions, creating computational artifacts, testing and refining computational artifacts, and communicating about computing). These concepts and practices showcase the wide scope of CS and the interrelatedness of CT. CS instruction should not be considered synonymous with programming. Moreover, although CT is also more than programming, programming can be an important tool for developing CT (Voogt et al., 2015). Elementary CS and CT instruction often emphasizes algorithms and programming with activities such as developing precise algorithms that computers can interpret, decomposing complex tasks into simpler tasks, and sequencing instructions either sequentially or by repeating patterns of instructions or by using events to initiate instructions K-12 Computer Science Framework, 2016. Although not the only computational concepts taught in elementary CS, algorithms and programming are often addressed as a means of practicing problem-solving.

CT integration

Another element that adds complexity for assessing CT in the elementary grades is that CT learning often occurs within an integrated context (e.g., along with science or mathematics learning). The rationale for integration includes (a) the natural "fit" of CT into certain disciplines (Lee et al., 2014; Weintrop et al., 2016), (b) opportunities to deepen students' understanding of content knowledge (Bers et al., 2014; Yadav et al., 2016), (c) limited time in elementary schedules for new content areas (English, 2017), and (d) promoting equity by providing CT within content areas taught to all learners (Israel et al., 2015; Santo et al., 2019). As part of a larger project, we have been developing curricular lessons that integrate CT within mathematics for students in Grades 3 and 4 (Strickland et al., 2021) and assessments to complement these lessons.

Given integrated approaches to CT instruction at the elementary level, CT assessments must be developed to measure CT learning goals while being cognizant of the specific instructional context in which CT is introduced. This challenge is not unique to CT education, as it often occurs when assessing learning in integrated science, technology, engineering, and mathematics (STEM) education (e.g., Douglas K. et al., 2020; Harwell et al., 2015). In this paper, we describe the approach we adopted that emphasizes assessing students' CT knowledge, skills, and abilities within the context of curricular lessons that integrate CT into mathematics. As such, we are interested in making claims about students' CT competencies; we are not trying to simultaneously measure both their CT and math knowledge, skills, and abilities by using the same assessments. Instead, we focus on developing assessments from the same CT foundation as used to develop lessons. Many of our CT assessments do involve mathematics, but do so to provide a context or a problem to be solved, rather than as a means to measure students' math competencies.

Learning trajectories to inform material development

At this time, there is no systematic, evidence-based means to decide *how* elementary students should learn CT (Zhang & Nouri, 2019). In the absence of such empirical evidence, our decision making relied on six learning trajectories (LTs) that were developed based on implicit goals in research studies in the areas of sequence, repetition, conditionals, decomposition, variables, and debugging (Rich et al., 2018, 2020, 2019, 2017).

LTs have three primary components: goals, developmental progressions of students' thinking processes, and instructional activities that promote the development of students' thinking along the progressions (Clements et al., 2020; Maloney et al., 2014). When LTs are used to guide lesson development, lessons are designed to build upon contiguous levels of instruction in a deliberate, consecutive manner because those lessons build on foundations set by previous lessons in order to avoid gaps in learning. If instruction is based solely on CT target competencies (e.g., lessons on sequencing), we assume that students will be able to perform tasks associated with that level of instruction as well as all previous levels.

The CT LTs used in this study were previously developed after analyzing CS/CT education peer-reviewed articles that were categorized by the learning goals identified in those studies. These learning goals were then organized into consensus goals (CGs) that included both an understanding goal (i.e., what students should know) and an action goal (i.e., what students should be able to do). These LTs were designed with the purpose of beginning to understand how students learn CT. As such, it is important to acknowledge that although these LTs may appear to be organized in a linear manner, learning does not occur in such a linear manner. Additionally, these LTs are considered hypothetical as empirical evidence for these LTs is only beginning to emerge as researchers begin to test students' CT development (e.g., Luo, 2020); doing so requires assessments that can be aligned to the LTs. Lastly, each LT does not exist in isolation; rather, they are connected. As students progress through one LT, progress on other LTs may be required or supported. For example, the sequence goal, *Computers require precise instructions using limited commands* can naturally lead into the repetition goal, *Instructions like "Step 3 times" do the same thing as "step, step, step"* and the decomposition goal, *Systems are made up of smaller parts*. These goals can lead into a conditional goal such as, *Actions often result from specific causes*. Thus, each LT is distinct, but increasing knowledge in one LT supports understanding in other LTs. More information about the development and

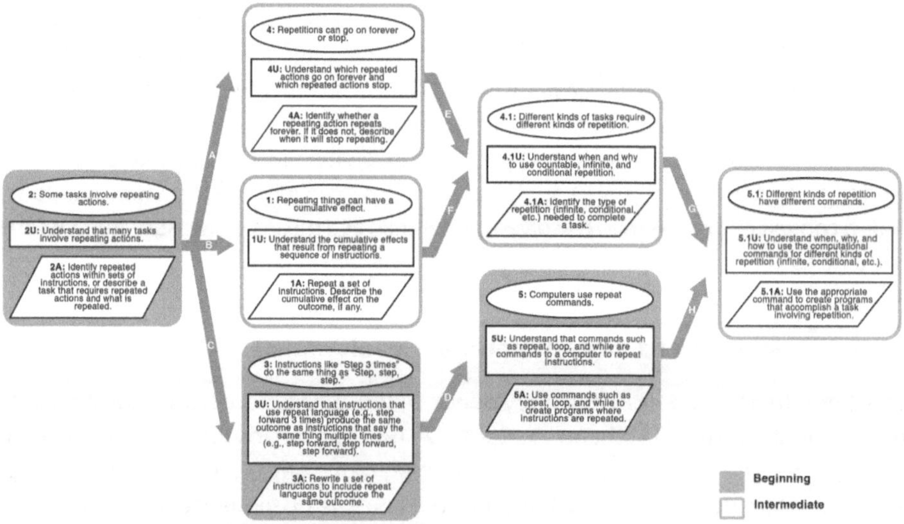

Figure 1. Example learning trajectory (LT): A portion of the full repetition LT, showing the statements that were used to develop the repetition design pattern. Shading is used to demarcate beggining goals (shaded) from intermediate goals (non-shaded), as defined by Rich and colleagues (2017); advanced goals are not included in this example. Each set of statements includes a consensus goal, an understanding goal (identified with an "U") and an action goal (identified with an "A").

features of the CT LTs are available in several papers by Rich et al. (2017), (2018), (2019), (2020) and on our website (http://everydaycomputing.org/).

Figure 1 depicts a portion of the repetition LT. This LT provides a series of learning goals for how elementary students can begin with a common understanding of repetition (i.e., *2: Some tasks involve repeating actions.*) as a starting point for understanding iteration in everyday life (i.e., *1: Repeating things can have a cumulative effect.*), and the utility of loops in programming (i.e., *5: computers use repeat commands.*). Each learning goal is connected to others to indicate progressions in knowledge, skills, and abilities. Our lessons were designed to guide students to build on each learning goal's level of thinking.

Curricular lesson development

Although the CT LTs extend beyond 3rd and 4th grade, our team focused on CT learning in 3rd and 4th grade fractions instruction, which were designed to guide learners in achieving the knowledge, skills, and abilities described in the six CT LTs. Thus, a series of 11 3rd grade and 12 4th grade lessons were developed to replace or supplement instruction available through the *Everyday Mathematics* (4th ed.; EM4) curriculum (Strickland et al., 2021). EM4 was chosen as it aligns to the Common Core State Standards for Mathematics (CCSS-M) and is a commonly used elementary curriculum in local school districts. Within the EM4 curriculum, we focused primarily on fractions instruction for two reasons. First, many students find proportional reasoning involving fractions difficult (Boyer & Levine, 2012). Second, CT and fractions instruction appear to be complementary. For example, many fraction standards in Grades 3 and 4 are framed around the notion that unit fractions can be combined repeatedly to produce non-unit fractions, which is related to the idea of repetition and looping in CT.

The 3rd and 4th grade lessons included a combination of unplugged activities (i.e., lessons that do not require interactions with the computer) and plugged activities (i.e., lessons that included computer-based activities). For plugged activities, we utilized the Scratch programming environment––a visual block-based programming environment designed for young learners Maloney et al., 2010––for several reasons. First, it is one of the best known programming environments for teaching CT to young learners (Price & Barnes, 2015). In fact, as of October, 2020, Scratch is the highest block-based programming language on the TIOBE index (https://www.tiobe.com/tiobe-index/), which measures search engine results of the popularity of programming languages. Second, Scratch is supported on Chromebooks, which were the computers available in our research schools. Most importantly, Scratch was chosen for pedagogical and theoretical reasons. Scratch was designed using a constructionist viewpoint that encourages young learners with no previous programming experience to explore, learn, and share Maloney et al., 2014. It was designed based on Papert's assertion that programming environments should have a "low floor" (i.e., easy to start) and a "high ceiling" (i.e., options to increase complexity; Resnick et al., 2009).

Regardless of whether the lessons were plugged or unplugged, lesson design used the same guided instruction approach aligned to the EM4 lesson design. All lessons started with a "warm up" that activated prior knowledge, made instructional learning goals explicit, and provided an introduction to the new content. This was followed by the

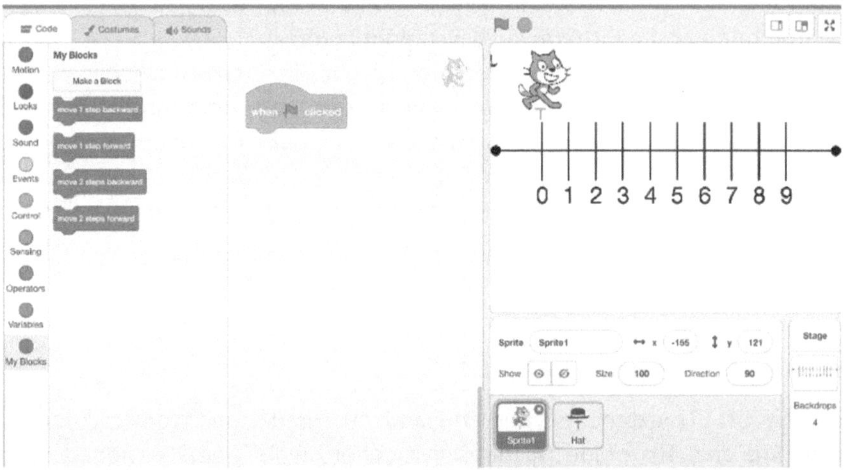

Using the blocks shown, create 2 different scripts (sets of instructions) to move the cat so that he stops at 5 on the number line. Write (or draw) your scripts in the boxes below.

<table>
<tr><th>Script A</th><th>Script B</th></tr>
<tr><td>Exemplar response*:
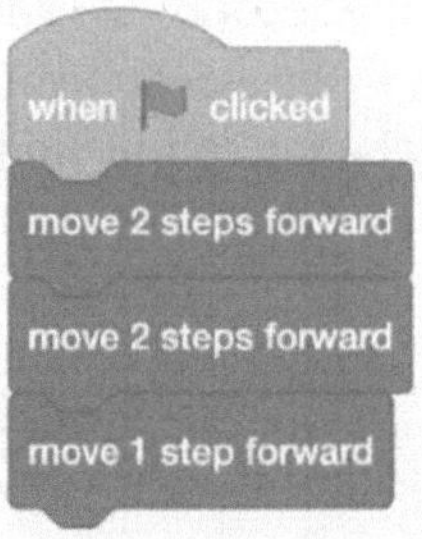</td><td>Exemplar response*:
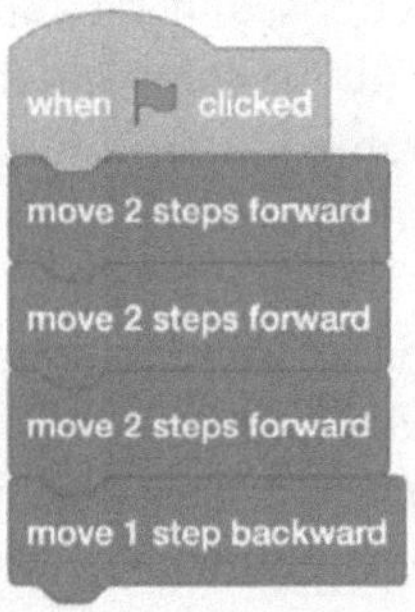</td></tr>
<tr><td colspan="2">*Any combination of forward and backward steps is acceptable as long as the cat stops at 5.</td></tr>
</table>

Figure 2. An example of an assessment item [S.01.a] that uses the Scratch interface and is aligned to the sequence LT. This version includes the exemplar responses which students would not see.

"focus" portion, the main instructional time, and ended with a wrap up to revisit goals and reflect on learning (Strickland et al., 2021).

Assessment development and field testing

As with the curricular lessons, we used the LTs to develop assessment items to measure students' CT knowledge and skills. There are three main features of the assessment items. First, all assessments used a "paper-and-pencil" format, so they did not require

Andre has 9 cookies that he wants to give away to 3 friends. He wants to give each friend an equal number of cookies. Write instructions for giving the cookies to the 3 friends. Make sure that each friend gets the same number of cookies. Use the instruction "repeat 3 times" at least one time.

Write your instructions below:

Exemplar response:

1. Repeat 3 times:
   a. Give first friend 1 cookie
   b. Give second friend 1 cookie
   c. Give third friend 1 cookie

[OR]

1. Repeat 3 times:
   a. Give first friend 1 cookie
2. Repeat 3 times:
   a. Give second friend 1 cookie
3. Repeat 3 times:
   a. Give third friend 1 cookie

Figure 3. An example of an assessment item [R.01.a] that does not use the Scratch interface and is aligned to the repetition LT. This version includes the exemplar responses which students would not see.

interactions on a computing device. Although they were paper-and-pencil, some assessment items used the Scratch interface. These items included illustrations of Scratch code blocks (and/or the Scratch interface) and/or asked students to draw Scratch code blocks by hand (e.g., Figure 2). Other assessment items were designed to elicit students' CT without being embedded within the Scratch programming environment (e.g., Figure 3). Second, some items used a mathematics context while others did not. However, the mathematics content was simplified for the assessment items to reduce the likelihood that students' proficiency with the focal mathematics content would preclude them from answering the items in a way that would demonstrate their CT proficiency. For instance, the item in Figure 2 asks students to move the cat along the number line which uses integers. In the lesson, students complete a similar task in which the number line uses fractions. By using integers in the assessment task, students without a full understanding of fractions could still complete this task (in the *Everyday Mathematics* curriculum, students have experience using a number line with integers in a prior grade). Third, all assessment items were designed as measures of a single LT topic (e.g., repetition). Thus, the produced assessment items could be classified along three dimensions:

(1) Whether the item does (or does not) use the Scratch interface,
(2) Whether the item is (or is not) embedded in a mathematics context, and
(3) Which LT (topic) is the primary focus for that item.

Román-González et al. (2019) classify CT assessment tools based on their method of evaluation: diagnostic tools, summative tools, formative–iterative tools, data-mining tools, skill transfer tools, perceptions–attitudes scales, and vocabulary assessment. Our assessments are designed as tools to measure near transfer that can be used diagnostically and/or summatively.

Theoretical and design frameworks

In this section, we describe the theoretical and design frameworks used to develop the assessments and interpret the assessment results. Underlying assessment use is the principle that the assessment must be valid with respect to the intended interpretive use. The design and interpretation of assessments are intimately intertwined and jointly affect valid use of assessments. Therefore, before describing the development approach and interpreting results, we frame our assessments within theoretical understandings of validity. We use the assessment triangle (Pellegrino et al., 2001) to structure the chain of reasoning between interpretation and use. To systematically apply the concepts represented by the assessment triangle, we use evidence-centered design. Each of these theories and design frameworks are described in turn.

Validity

In their systematic review of CT assessments, Tang and colleagues (Tang et al., 2020) conclude that across almost 100 studies, only 18% reported on the validity of their assessments. Those that did consider validity tended to focus on criterion, content, and construct validity. In this paper, we follow a different tack, focusing on contemporary approaches that conceptualize validity as an argument (AERA, APA & NCME, 2014; Douglas & Purzer, 2015). In an argument-based approach to validity (Kane, 1992, 2006), one makes claims about the intended interpretive use of the assessment and then provides evidence in support of those claims. This evidence can and should be multifaceted, including information about the design of the assessments as well as empirical evidence from student performance (Crocker, 2006; Pellegrino et al., 2016). After instruction with the curricular lessons, students should be able to demonstrate CT as they solve items that were designed to recruit specific aspects of CT. Specifically, they should be able to use the knowledge, skills, and abilities that are developed during instruction and enactment of the lessons.

For assessments that are intended to support instruction, there are three components of validity that can be considered: instructional, inferential, and cognitive; each have their own associated claims and potential sources of evidence (Pellegrino et al., 2016). This paper focuses on providing preliminary evidence for claims related to the cognitive and inferential components of validity. The cognitive component is centered on the assessment items being an accurate measure of students' CT as articulated by the LTs, and in this paper is evidenced by the assessment design methodology. The inferential component is related to whether the assessments reliably differentiate among students and support model-based analyses of their measurement properties and in this paper is evidenced by results from the analysis of students' performance on the assessments.

We have one primary claim that we advance and support in this paper, namely that **the CT assessment items measure students' performance with respect to the LTs (topics) that students were introduced to during the lessons**. This primary claim has three constituent subclaims:

(A) Items are aligned to the LTs and students' opportunities to learn,
(B) Item scores are reliable (i.e., item scoring rules and rubrics can be used by different people to arrive at the same conclusions about students' performance), and
(C) Items are appropriate for learners within the target grades who have varied levels of proficiency with the LTs (topics).

In this paper we will provide evidence for the three subclaims and thus for the primary claim about the appropriateness of the CT items. In the discussion section we return to an evaluation of these claims and look ahead to other claims that we would like to evaluate in future work.

Assessment triangle

The assessment triangle is a framework for representing assessment as a process of reasoning from evidence, focused on three interconnected elements for assessment design and interpretation: cognition, observation, and interpretation (Pellegrino et al., 2001). The cognition vertex refers to cognitive theories and models of knowledge representation, processing, and development in the domain (e.g., CT). The observation vertex refers to the design of assessments that will provide evidence of what students know and can do. As such, it is guided by elements of the cognition vertex. The interpretation vertex refers to how one makes reasoned inferences from the collected observations; it can include both rubrics and statistical models. All three vertices should be considered individually, and in combination with each other (Pellegrino et al., 2001). The assessment triangle, as a framework, can be used to frame how one designs assessments and interprets the results of those assessments (e.g., Streveler et al., 2011).

Evidence-Centered Design (ECD)

Evidence-centered design (ECD) is a framework and a design process that considers assessment design across multiple layers of implementation (Mislevy et al., 2003). While the assessment triangle is a theoretical framework, ECD provides steps and processes to ensure that assessment developers focus on all three vertices of the assessment triangle. Although often used for large-scale assessment design, ECD can also be used to design classroom-based assessments. Fundamentally,, ECD represents a space of claims about what students know and can do, an articulation of the evidence that students could provide (in what they say or do) that could be used as evidence for these competencies, and the features of assessment items that could elicit such evidence (Pellegrino et al., 2014). The present research and development process used ECD, most prominently by creating *design patterns* that describe families of assessment items (Mislevy & Haertel, 2006). These design patterns are critical, in part because they define the *knowledge, skills, and abilities* (KSAs) that underlie each of the different CT LTs. In essence, the KSAs become the claimed competencies for which one attempts to elicit evidence through the design of assessment items with certain features (Pellegrino et al., 2014).

The Principled Assessment of Computational Thinking (PACT) project has also used ECD to develop CT assessments. Bienkowski and colleagues (Bienkowski et al., 2015) developed four design patterns, each representing a different CT practice (e.g., *Design and implement creative solutions and artifacts*). These design patterns were independent of a specific curriculum, programming language, or grade level. Because of this independence, they allow for reuse across different assessment contexts. These design patterns were then used to develop assessments, for instance, for HS students in an introductory CS course (Snow et al., 2019). One key distinction between the PACT ECD work and ours is that the PACT group conducted their own domain analysis, whereas our domain analysis was effectively completed through the previous efforts to develop the LTs (Rich et al., 2018, 2020, 2019, 2017). Those LTs were also critical in informing and "jump-starting" the domain modeling process as they directly fed into our design patterns.

Moving from learning trajectories to assessments

The following sections describe how we used a principled approach to work forward from the LTs to develop the assessment items, using the assessment triangle to frame our description.

Cognition vertex

Our cognition vertex is based on the CT LTs (e.g., Figure 1). As mentioned earlier, the LT learning goals include three statements: an understanding (U) goal, a student action (A) goal, and a summary/consensus goal that describe the overall learning goal. We started by aggregating the learning goals within an LT and then used that set of statements to develop a design pattern for the LT. This procedure is elaborated in the next section.

Observation vertex

Design patterns. We used a knowledge representation called a design pattern that assists in specifying key details of the assessment argument, ensuring that developers are explicit in identifying the claims, evidence for those claims, and methods by which one may elicit that evidence (Mislevy & Haertel, 2006). We created design patterns for five of the six CT trajectories: decomposition, sequence, repetition, conditionals, and variables.[1] Each design pattern was then used to develop a family of assessment items that individually and collectively could elicit evidence that students possessed the knowledge, skill, and abilities associated with respective LT.

There are four main categories of design pattern components: description (title, summary, rationale), KSAs (focal and additional), evidence statements (potential observations and potential work products), and item features (characteristic and variable). To illustrate our process for creating the design patterns we describe these four categories and how the LTs informed their development. We use the repetition LT (Figure 1) and the repetition design pattern (see the Supplementary) as an example to illustrate the procedure that was used for all five design patterns. Two design pattern features are worth noting. First, connections to the source LT statements are noted with alphanumeric codes (e.g., "3; 3 U"). Interested readers can use these codes to trace content in the design pattern to its source in the LT (i.e., Figure 1). Second, the design pattern uses a convention in which occasional text is struck (e.g., "forever"); this is used when a LT statement included content that was not covered by the curricular lessons (e.g., in the lessons students did not learn about *forever* loops but did learn about *repeat until* and *repeat X times* loops). Strikethrough text was not used to develop assessment items.

The description category of the LT provides a brief summary and rationale for the topic that is represented in the design pattern. We wrote the description components to capture the full range of learning goals expressed in the LT, occasionally using language from specific learning goals (e.g., 5A in the repetition design pattern summary).

The knowledge, skills, and abilities (KSAs) category describes the cognitive foundations necessary for successfully responding to items written from this design pattern. Identifying and defining focal KSAs allows one to articulate the multiple proficiencies that form the design pattern and thus measurement targets; these focal KSAs can also be used to develop an ECD student model (Bienkowski et al., 2015; Mislevy et al., 2003). We used CT LT statements to define focal KSAs. We combined, reordered, and collapsed statements when needed, in the process rewording them. For example, the focal KSA *"Knowledge that different kinds of tasks require different kinds of repeated instructions and therefore different repeat commands"* is a combination of LT statements 1, 4.1, 4.1 U, 5.1, and 5.1 U (see the Supplementary and Figure 1). Additional KSAs represent additional cognitive factors that might be required to respond to items, but are not measurement targets. This step of defining KSAs most clearly illustrates the hypothesized cognitive connections between the assessment items and the LT.

Design patterns also include potential observations and potential work products. Once KSAs are articulated, one can define the potential observations and potential work products that would provide evidence for whether a student can use those KSAs. Collectively the potential work products and potential observations are evidence statements--descriptions of what students could say or do that would count as objective evidence that they have the competencies claimed by the design pattern (Pellegrino et al., 2014). As with the KSAs, we used the LT statements to help write evidence statements, along with considerations about how students could demonstrate the KSAs. Often, the student action goal was useful in writing the potential observations (e.g., 4.1A).

Finally, design patterns include characteristic item features and variable item features. Together, the potential observations and student work products suggest and constrain potential features of assessment items that might elicit the intended evidence. Characteristic item features define the features that all items developed from the design pattern will have. In contrast, a variable item feature defines a feature that some items will have, or a dimension along which to vary items. When creating the design patterns, some of the item features were written based on a LT action goal while others were written from the KSAs. One variable feature present in all five design patterns was the presence of a mathematics application/context. A subsidiary characteristic feature was that such items should use whole numbers, not fractions. In this way, we connected to the curricular lessons, but ensured that those students who were still developing their fraction understanding would not be precluded from demonstrating their CT. We used this approach to try to minimize construct-irrelevant variance (AERA, APA & NCME, 2014).

In summary, the LT statements were used to write claimed competencies articulated in the form of focal KSAs. These KSAs were used (in concert with other LT statements) to develop the remainder of the design pattern (i.e., evidence statements and task features). These design pattern components were in turn used to guide the development of the assessment items. This allowed us to use the LTs to define the cognition vertex of the assessment triangle, and to use ECD to link the cognition and observation vertices.

Developing Assessment Items. We used the design patterns to develop individual assessment items that might elicit one or more focal KSAs. Potential observations and work products were used to suggest item designs, and the characteristic and variable features were used to suggest how to create and vary items from the same design pattern. In doing so, we used a variety of response formats, including different types of selected and constructed response formats. Some items used the Scratch interface and/or code blocks and others did not. Because design patterns define a family of assessment items, the developed items are not exhaustive; future developers can use the design patterns to develop additional items that align to the LTs.

As part of the item development process all items included an *exemplar student response* that signaled what a student response would look like for a student that had the relevant KSAs. The focal KSAs and evidence statements were used to guide and check the exemplar student responses. Before asking students to complete the assessment items, all items were reviewed by project members with experience teaching CT and writing CT curriculum for the target grade levels. These reviews focused on item clarity, language, accessibility, and alignment to the learning goals. Items were revised following review.

Developing Assessment Instruments. The next step was to assemble individual items into instruments. We created six instruments. Each grade had three instruments that were designed to be administered after specific lessons were enacted (the assessments were administered after completing sets of four lessons and were labeled according to the administration chronology, i.e., Early, Mid, and Late). This allowed us to select items to match the content that was being learned at the corresponding time. Because the content of the lessons constrained the relevant assessment items, a single instrument often had different numbers of items for each LT and included items from only some LTs. Attempting to include assessment items from all LTs in equal number on each instrument would have created misalignment between the assessments and students' opportunities to learn the content.

Each instrument sampled from the content of the different LTs that students had been taught in the preceding curricular lessons, included approximately 10 assessment items, and was designed to be completed in one class period. We selected items in approximate correspondence to the distribution of LT goals throughout the associated lessons. For example, the Grade 3 Early assessment had decomposition items (n = 5), sequence items (n = 3), and repetition items (n = 2), while the Grade 4 Early assessment had decomposition items (n = 3), sequence items (n = 3), and variables items (n = 4). Forty-four items were used across the six instruments. Although some items were used on multiple instruments, the majority of items appeared on only one instrument. This allowed us to maximize the number of items that we could collect data for, while limiting the length of each assessment instrument.

Pilot testing

During the 2018–2019 academic year, we piloted the curriculum and assessment instruments. Partly because this was a pilot year for curriculum and assessment use, data collection was less complete than planned. As the academic year progressed, the number of teachers that used the assessment instruments decreased due to constraints on classroom time. We were able to collect data on five of the six assessment instruments (all except the Grade 4 Late); of these five, the two Early assessments (Grade 3 and Grade 4) were used the most widely. Therefore, in this paper we focus on these two instruments because they have the most student data. The analytic approach we describe provides an initial validity evidence and illustrates the approach that could be used with the other four instruments. As elaborated in the Discussion section, this paper describes only a portion of the larger assessment use and validation approaches that we are engaged in to develop LT-based CT assessments for upper elementary students.

Context and participants

Data were collected from students in five elementary schools (four public schools; one private school) in two midwestern states in the U.S. with diverse student populations. The school-level demographic data for each school is reported in Table 1. For the four public schools where state reports were available, each had a diversity in cultural backgrounds, socioeconomic status (i.e., from low income families or needing free/reduced price meals), educational needs (i.e., enrolled in special education), and language background (i.e., English language learners). Five Grade 3 classes and three Grade 4 classes had students

Table 1. Demographic data for participating schools (2018–2019 academic year).

School	Ethnicity	Students with IEPs (%)	Low Income Students (%)	English Learners (%)	Grade 3 Classes	Grade 4 Classes
A	40.2% White, 30.9% Black, 9.8% Asian, 6.2% Hispanic, 12.9% Two or More Races	19	59	7	0	1
B	50.6% White, 19.9% Black, 13.7% Asian, 7.9% Hispanic, 7.9% Two or More Races	13	43	13	1	0
C	51.2% White, 20.4% Black, 11.2% Asian, 7.2% Hispanic, 10.1% Two or More Races	13	31	9	1	0
D	62.1% White, 28.2% Hispanic, 2.9% Black, 2.4% Asian, 0.2% American Indian, 4.2% Multicultural	19	43	19	1	1
E	[no data]	[no data]	[no data]	[no data]	2	1

Source: Illinois Report Card (https://www.illinoisreportcard.com/) & Indiana Department of Education.

Table 2. Example rubric for assessment task (R.01.a).

Score	Description	Examples of student responses
2	Instructions use the command "repeat 3 times" and produces the intended outcome.*	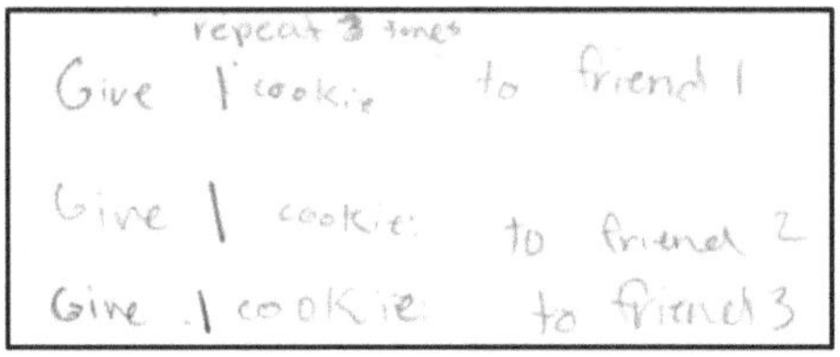
1	Shows understanding of repetition to get desired result through word explanation or drawings, but instructions won't produce the intended outcome.*	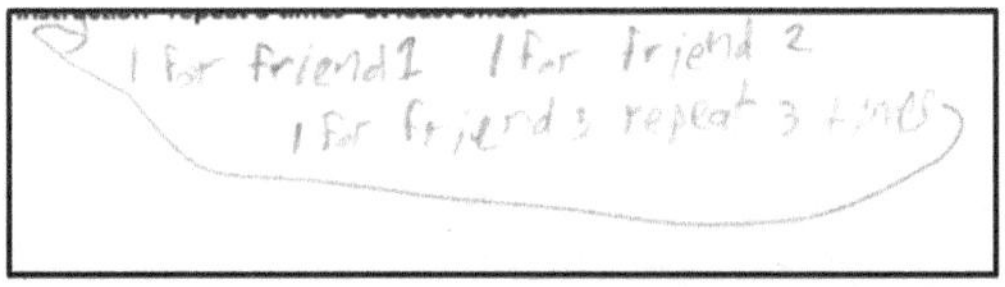
0	Incorrect use of repetition concept, or no demonstration of repetition concept to achieve the intended outcome.	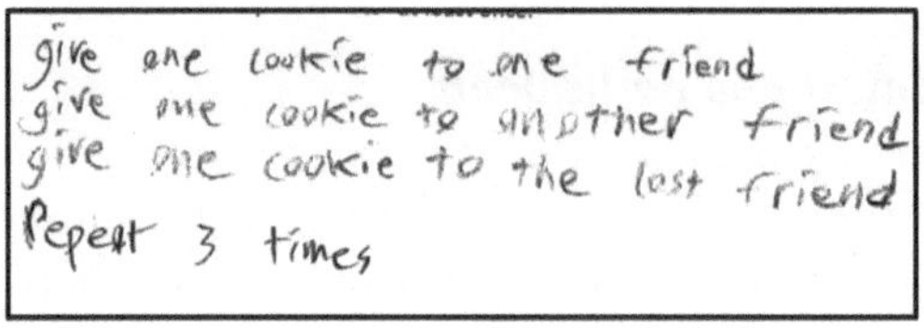 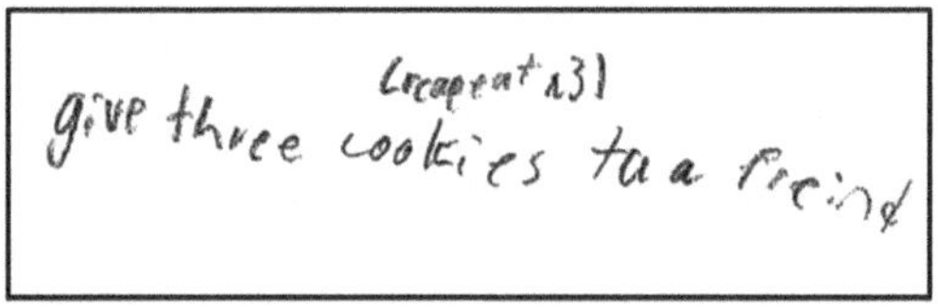 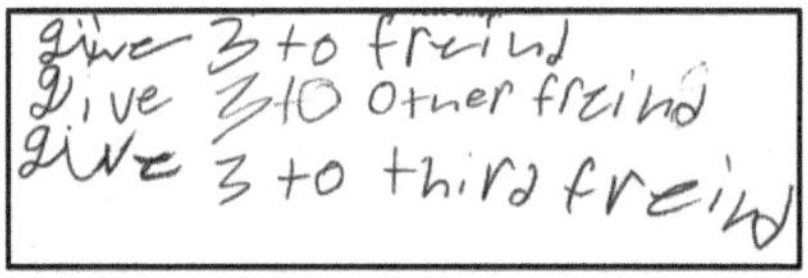 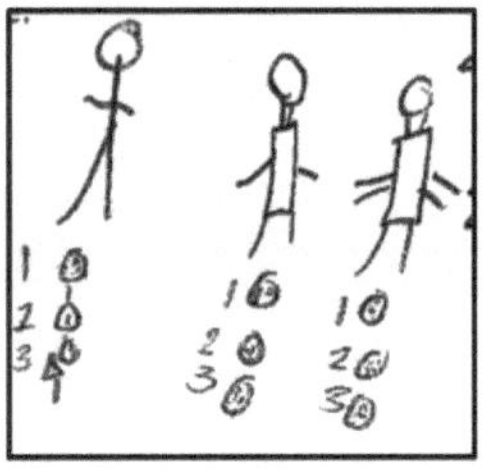 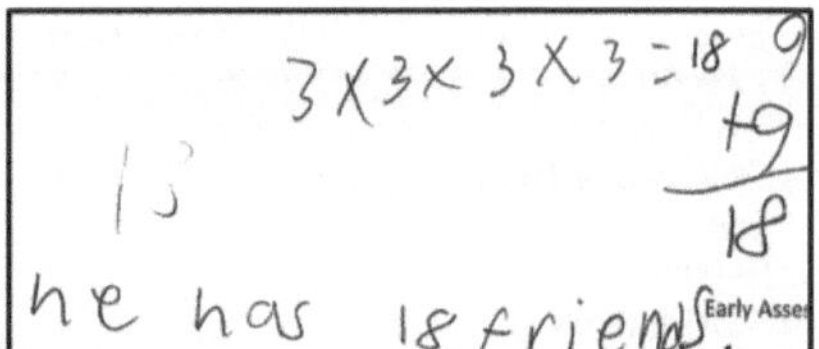 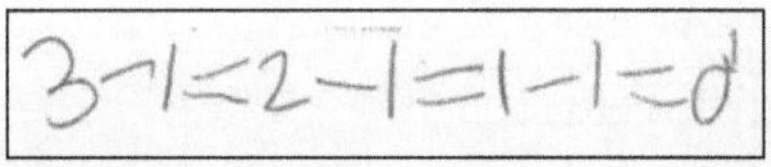

*Responses with drawings must demonstrate repetition to achieve the intended outcome

that completed the assessment instruments. We obtained parental consent and assent for students in the participating classes for data collection. The assessment instruments were administered after the teachers (with researcher/curriculum developers' assistance) enacted the integrated lessons. As this was our first year to pilot both the lessons and assessments, students in Grade 4 had not received prior Grade 3 instruction as would be expected in a full implementation of the curricular lessons.

Scoring (interpretation vertex)

First, we determined how to interpret students' responses using objective criteria. We coded student responses for accuracy and characteristics of their responses. Each item's exemplar response was used to signal what a "correct answer" was, or what feature(s) a correct answer would have. The majority of items had simple scoring criteria that corresponded to a correct answer for the selected response items and for some constructed response items. When items resulted in a range of student responses, rubrics were developed to aid scoring. Rubrics were developed after reviewing a subset of student responses and considering the type of knowledge and skills that the assessment items were designed to elicit. This process resulted in different categories of student responses that were then coded using the rubric to determine a single score. An example rubric (for the item shown in Figure 3) is provided in Table 2. This rubric has three categories (scores) that correspond to different levels of a student response that might show evidence of using the repetition construct. Student responses were identified that would typify responses for each category, and added to the rubric as examples. In the first category (score = 0) the student might have illustrated the correct outcome, but without using repetition (e.g., they designate each of the friends should receive three cookies, but they do not use any repetition to do it). In this case, because the assessment item is aligned to the repetition trajectory, it is explicitly aimed at providing students an opportunity to demonstrate they can use repetition; providing an alternate method to achieve the correct outcome, without using repetition, does not provide evidence that that student can write instructions using repetition. The item was designed to elicit this ability (the item instructs students to "Use the instruction 'repeat 3 times' at least one time."). In the second category (score = 1) the student might have shown they understood that an instruction/command could be issued once and then repeated (perhaps multiple times), however, the student's instruction does not achieve the intended outcome. In contrast, in the third category (score = 2), the student both achieves the intended outcome and gives an indication that that an instruction/command is to be repeated.

To ensure reliable coding of students' responses we used a dual-rater method with 100% overlap of dual-coded responses. This method involved an initial training session, followed by coding a subset of the data, an interim check on agreement, and then a complete coding of the data. In the initial training session both coders and the lead author collectively applied the scoring criteria and rubrics to three students' responses. During this session adjustments and clarifications to the scoring criteria or rubrics were made as needed. The two coders then independently scored approximately 30% of the responses. Percent agreement was calculated and disagreements were resolved through discussion between the two coders, moderated by the lead author; particular attention was paid to any items where agreement was below 80%. During these discussions we revised the scoring criteria or rubrics as needed to clarify scoring decisions. Following this

Table 3. Item parameters for the Grade 3 Early assessment instrument (after dropping Item DC.04.a).

Item	Difficulty (Mean)	Variance	Discrimination
DC.02.a	.72	.45	.47
DC.03.b	.79	.41	.40
DC.05.a	.74	.44	.40
DC.08.a	.62	.49	.63
S.01.a	.52	.50	.87
S.04.b	.60	.49	.87
S.06.a	.66	.48	.47
R.01.a	.24	.43	.53
R.05.a	.57	.50	.90

interim check the two coders scored the remaining 70% of the data. Again percent agreement was calculated and disagreements were resolved through discussion; scoring criteria and rubrics were updated as needed to reflect the scoring decisions that were used or finalized during that final session.

We report reliability using Cohen's Kappa (weighted) which can range between −1 and 1, with a higher value indicating greater agreement among raters. Weighted Kappa allows one to compute agreement among ordinal scores while allowing scores that are closer to yield higher agreement (Viera & Garrett, 2005). We computed Cohen's Kappa for each item (excluding the three students' responses that were used for each training session), for each assessment instrument. Overall, we had high reliability on both the Grade 3 Early instrument (median = .98, min = .94, max = 1) and the Grade 4 Early instrument (median = .91, min = .55, max = 1).

Results and findings

We conducted a multi-step analysis of the student response data, analyzing each grade's instrument separately. When reporting these results we reference the item codes to identify items. Our item code naming convention denotes the trajectory (e.g., "DC" = decomposition, "S" = sequence, etc.) and the item number (e.g., "01", "02", etc.), and uses a final indicator to distinguish between variations in similar items (e.g., "a", "b", etc.).

Grade 3 Early assessment

Ninety students from five classes (four teachers; one teacher taught two classes) completed the Grade 3 assessment following instruction. Before data analysis, we dropped one of the decomposition items (DC.04.a) because of concerns raised by project members that it was not measuring any of the decomposition KSAs. The remaining nine items sampled from items developed to align to one of three design patterns/LTs: decomposition, sequence, and repetition. After initial data exploration, we dichotomized the three items (DC.08.a, S.01.a, R.01.a) that had been coded polytomously (i.e., using partial-credit scoring as illustrated by the rubric in Table 2). For each item scored polytomously, we collapsed coding across score categories based on the item, rubric levels, and distribution of student responses among rubric levels. For instance, continuing with the R.01.a

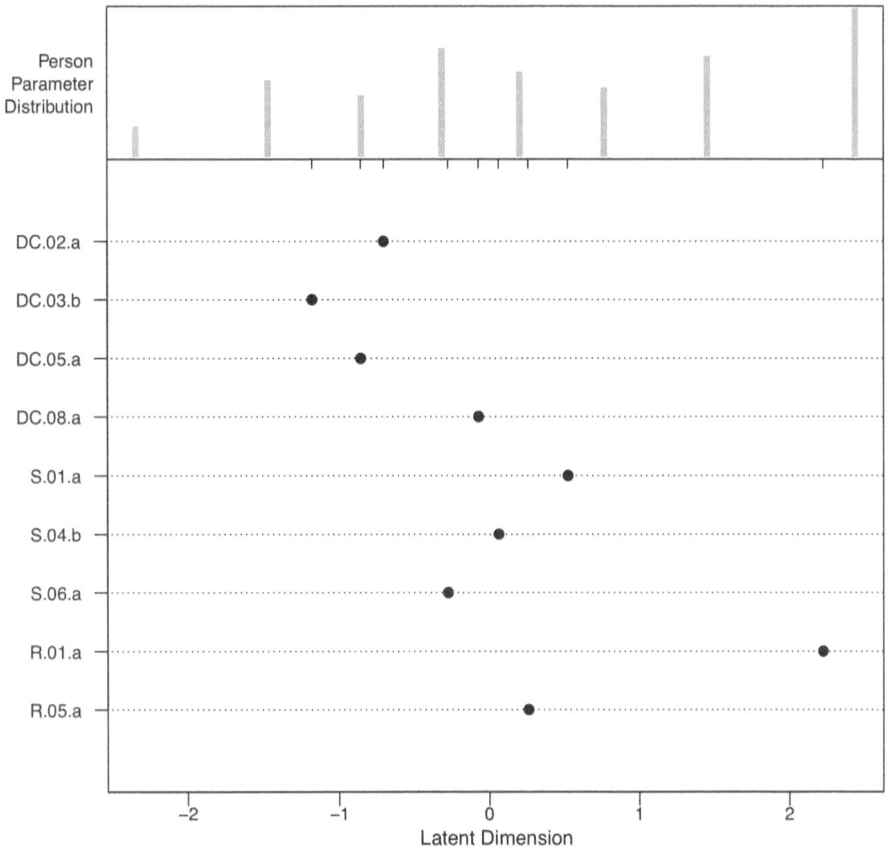

Figure 4. Person-Item map (akin to a Wright map) that shows the joint distribution of item difficulties (thresholds; lower panel) and person ability estimates (upper panel) for the Grade 3 Early assessment instrument. The upper panel displays a histogram of student ability estimates. The lower panel contains a row for each item, with its associated difficulty indicated with a point.

example (Figure 3 and Table 2), we collapsed two categories (where score = 1 or score = 2) into a single category (score = 1) and retained the remaining category (where score = 0).

To understand how the items were performing individually, we conducted classical test theory (CTT) analysis. Difficulty (*p*) values indicate the "easiness" of an item and so (counterintuitively) higher difficulty values indicate easier items and lower values indicate harder items (Crocker, 2006). For dichotomous items, difficulty values are the proportion of correct responses. Item R.01.a was the most challenging item for our sample of students to answer. Item DC.03.b, a multiple-choice item about why one should decompose an item, was the easiest item on the instrument with almost 80% of students endorsing a correct answer. All item difficulty values fell within the .20–.80 range of acceptability (see Table 3). We also calculated discrimination, a measure of how well an item can discriminate between students of "high" and "low" proficiency on the construct of interest. A higher discrimination value indicates that the item is better able to discriminate between these levels; a low (or negative) discrimination value indicates there might be problems with the item. All items had discrimination values greater than .20, a minimum value typically used as a cut-off to signal "acceptable" items (see Table 3).

To further understand these data, we conducted analyses using Item Response Theory (IRT) using the eRm package (Mair & Hatzinger, 2007) in R. IRT (Baker, 2001; Embretson & Reise, 2000) allows one to model students and items on the same latent trait (θ), enabling one to consider both item and student (person) characteristics simultaneously (e.g., Figure 4). Although a small sample, we fit a Rasch model to the data. An overall goodness of fit test (Anderson LR-test, $X^2(6) = 11.14$, $p = .08$), indicated the Rasch model fit the data. Figure 4 plots both person and item characteristics, notably ability (persons) and difficulty (items). The top quarter of the plot shows a histogram of person ability as estimated for students in our sample, where the histogram bars indicate the count and location on the latent trait scale. The remainder of the plot shows items' locations on the latent trait scale. Each point indicates where the item falls relative to a 50% probability that students would answer the item correctly as predicted on the basis of their overall ability estimate (based on their performance on the instrument as a whole). Item difficulty (β) showed a good range of measurement along the latent trait, with values ranging from −1.18 to +2.23 (see Figure 4). Although there were a small number of items, the items show a range of difficulties, a desirable feature for an assessment that will be used with students with a range of proficiencies on the CT construct.

Overall, students' performance on the assessment instrument was moderate. Students answered more than half of the nine items correctly ($M = 5.5$, $SD = 2.5$). As indicated earlier, this is a desirable finding for collecting initial evidence of the appropriateness of the instrument for use with students that have a varying range of abilities. For potential diagnostic and formative use one does not want a test that is too easy or too hard because it can limit the utility of the information provided by the assessment. Further, considering these items in light of the LTs from which they were designed, it appears that the decomposition items used on this instrument were the easiest items, followed by the sequence items, followed by the repetition items (as indicated by the difficulty estimates provided by both the CTT and IRT analysis).

Grade 4 early assessment

Fifty-four students from three classes (three teachers) completed the Grade 4 Early assessment following instruction. Before analysis we dropped the DC.04.a item for the same reason we dropped it from the Grade 3 assessment. We also dropped a variables item (V.07.c) because of a typographical error that created ambiguity around what students were expected to do to answer the item correctly. After dropping Items DC.04.a and V.07.c, eight

Table 4. Item parameters for the Grade 4 Early assessment instrument (after dropping Items DC.04.a, V.07.c, V.10.a, and S.06.b).

Item	Difficulty (Mean)	Variance	Discrimination
V.03.b	.35	.48	.44
V.04.c	.85	.36	.22
S.02.a	.31	.47	.77
S.04.d	.57	.50	.72
DC.02.b	.87	.34	.33
DC.03.b	.81	.39	.39

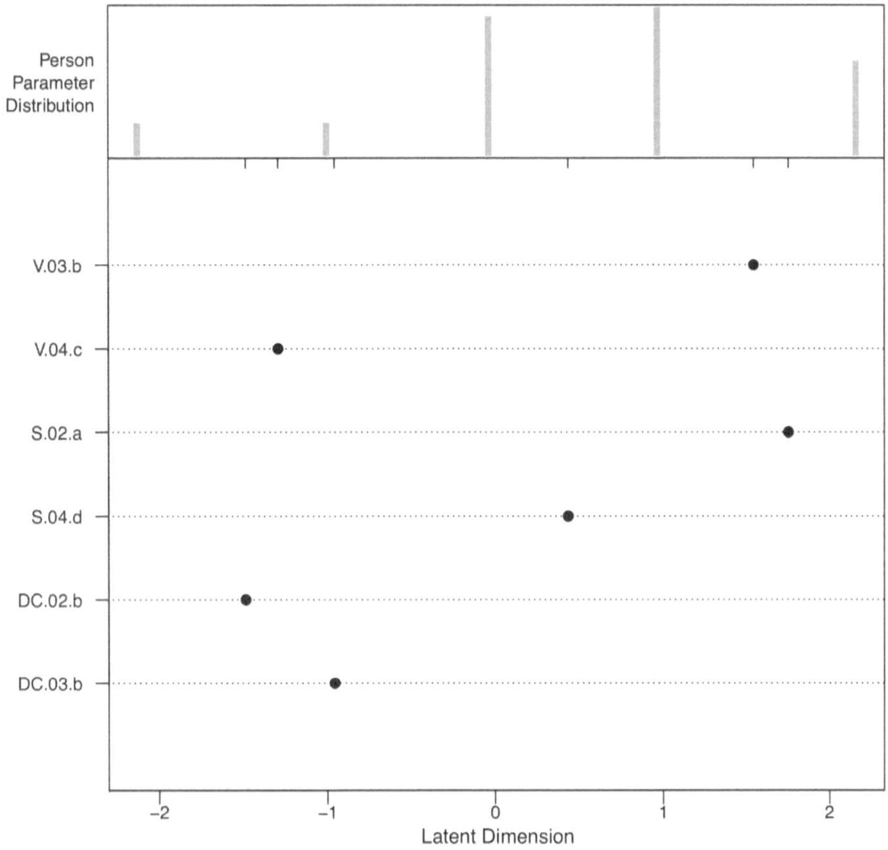

Figure 5. Person-Item map for the Grade 4 Early assessment instrument.

items remained. These items sampled from the variables, sequence, and decomposition LTs. After initial data exploration, we recoded one item (S.02.a) from polytomous to dichotomous. Additional data exploration involved looking at measures of internal consistency and correlations among items and total score. During this exploration we identified four potentially problematic items and examined the items and scoring criteria to understand whether the items might need to be removed before further data analysis. This investigation did signal problems with two of those four items. Item V.10.a used a multiple-choice format where students were asked to circle multiple correct answers (this format was potentially difficult because the instructions about selecting multiple responses were unclear). Perhaps more significant though, the content that the item was designed to assess was not reflected well by the content in the lessons. Item S.06.b used a true/false format. Upon review we determined that it used a statement that––although true in most cases––could reasonably be argued as false by someone experienced with Scratch. Because Items V.10.a and S.06.b were negatively correlated with total score and their review identified potential problems, we decided to drop them before conducting further analysis.

Difficulty and discrimination values for the six items that remained are reported in Table 4. Discrimination values were all acceptable. Three items had difficulty values greater than .80, indicating a large proportion of students got them correct and that they were easy for most students in our sample. This finding is further explored in the IRT results when we compare the items to students' ability estimates.

Although we had only a small sample of student responses, we fit a Rasch model to the data (Anderson LR-test $X^2(3) = 7.09$, $p = .07$). In calculating overall model fit two items had to be excluded due to their student response patterns (which is in part due to their easiness which resulted in little variance among subgroups). IRT difficulty showed a good range of measurement along the latent trait, with values ranging from −1.48 to +1.54. As with the Grade 3 Early instrument, items were spaced out along the latent dimension, indicating that some items (e.g., DC.02.b and V.04.c) can be answered correctly even by students that are estimated to have lower CT proficiency, and some are only likely to be answered correctly by those that are estimated to have a higher CT proficiency (V.03.b and S.02.a) (see Figure 5).

As with the Grade 3 Early instrument, students' performance on the Grade 4 Early assessment was moderate. On average, students answered more than half of the six items correctly ($M = 3.8$, $SD = 1.4$). Unlike the results for the Grade 3 Early instrument, there did not seem to be any clustering of item difficulty by LT, although both decomposition items were among the easier items on the assessment. With this sample, the two sequence items were the most highly discriminating (see Table 4).

Discussion

In this paper, we have shown how the three vertices of the assessment triangle were defined and coordinated, using ECD, to yield specific claims regarding the validity of CT assessments for use with upper elementary students. Our development methodology constitutes evidence for one part of our validity argument (Crocker, 2006; Pellegrino et al., 2016), and the empirical results constitute additional evidence for other parts of the validity argument.

Assessment development

Validation requires more than the application of psychometric procedures (Douglas & Purzer, 2015). It requires a principled design process that documents design decisions and collects multiple sources of evidence to support the proposed interpretation and use of assessment results (Pellegrino et al., 2016; Gane et al., 2018). We presented initial, positive evidence to demonstrate that one can use LTs to design assessments in a principled way and use those assessments to collect evidence that students can draw upon their CT knowledge, skills, and abilities to respond appropriately to the assessment tasks following instruction. The LTs provided a cognitive model, crucial for implementing the other parts of the assessment triangle (the observation and interpretation vertices), as well as for guiding the assembly of evidence regarding validity. These key aspects of design and validation have been notably absent from many prior attempts to develop CT assessments (Tang et al., 2020).

By explicating our assessment development method, we traced how we worked forward from the LTs, to design patterns, to assessment items, and finally to assessment instruments. In doing so, we highlighted critical details that are needed for a comprehensive validity argument.

Empirical results

Since this was our first round of implementation and data collection, it constitutes a pilot evaluation of both the lessons and the assessments. The number of participating students and teachers was not large, and this paper only reports on two of the six assessment instruments that were developed. As such, there are limitations on the interpretations and claims we can make given the empirical data. Despite those constraints, we were able to show that the two early assessment instruments and associated items had acceptable levels of interrater reliability. Our analysis of item and test performance signaled a small number of items that needed revision or exclusion, which we completed before more recent data collection efforts.

For the majority of items, CTT and IRT analyses provided evidence that the items were appropriate for students with a range of CT proficiencies. The items showed a range of difficulties, and acceptable levels of discrimination. Students performed reasonably well on the items as a whole, consistent with the conclusion that students have the knowledge and skills needed to answer these questions after instruction with the lessons. If the items had been too difficult for students, they would have appeared further to the right in the Person-Item maps in Figures 4 and Figures 5 (e.g., $\theta > 2$). Although these performance data suggest students learned from the lessons, we do not have the counterfactual (students with no CT instruction). Therefore, we do not know how students would have performed if they had not completed the curricular lessons.

Reflecting on the assessment item claims and evidence for these claims

As stated previously, we have one primary claim that we support in this paper: the CT assessment items measure students' performance with respect to the LTs (topics) that students were introduced to during the lessons. This primary claim involves three constituent subclaims, for which we have provided evidence in this paper. **Subclaim A (items are aligned to the LTs and students' opportunities to learn)** was supported by the evidence we have provided related to our process and our documentation of this process during development. In particular, we used a principled approach to convert the LTs into design patterns that were then used to write assessment items. These assessment items were reviewed by individual project members with experience teaching CT and writing CT curriculum for the target grade levels and then revised following review. Finally, assembling the items into instruments was done using the curriculum developers' mapping between lessons and LT goals to ensure that the content used to design the lesson activities was represented in the items that appeared on the assessment instruments, and that the items on the instrument did not go beyond the content used in the lessons. In this way, we were able to align items to both the LTs and students' opportunities to learn. **Subclaim B (item scores are reliable)** is supported by the evidence we have provided that the item scoring rules and rubrics can be used by different people to arrive at the same conclusions about students' performance. In particular, the interrater reliability results showed high agreement among scorers for all items and across both instruments. In addition, the process we used to develop and modify the scoring rules and rubrics during the training and subsequent use add to the evidence supporting our claim that the scores can be reliably determined. **Subclaim C (items are appropriate for learners within the target grades who have varied levels of proficiency with the LTs/topics)** is supported by the empirical evidence on the difficulty and discrimination of the items, along with the comparison between item difficulty and student proficiency (i.e., Figures 4 and Figures 5).

Although we are still engaged in assessment development, we present initial evidence for our primary claim and subclaims. By evaluating the primary claim we also provide support for the cognitive and inferential components of validity (Pellegrino et al., 2016). Subclaim A is most related to the cognitive component of validity. Evidence for Subclaim A applies to all the items that were created from the LT design patterns, because the same process was used for items across the five LTs and six instruments; the design evidence applies to even items that were not piloted. Subclaims B & C are most related to the inferential components of validity. For Subclaim B and C, the evidence we have in support of these subclaims is limited to those items used on the Grade 3 and Grade 4 Early instruments. Eventually, our goal is to provide evidence in support for these two subclaims for all items across the five LTs and six instruments.

Future directions

Validation is an ongoing process (Douglas & Purzer, 2015) that requires accumulating multiple sources of evidence tailored to the specific claims being advanced (AERA, APA & NCME, 2014; Pellegrino et al., 2016). Doing so requires multiple methodologies, sources of data, and types of analysis. As such, this paper presents an initial assembly of validity evidence and is not the full extent of our efforts to validate and use these assessments. For instance, we have also conducted cognitive interviews with students to further understand the knowledge and skills elicited by the assessment items relative to the targeted components of the specific LTs (Luo, 2020; Luo et al., 2020). In brief, these interviews helped identify items that needed revisions (e.g., to minimize the possibility that math knowledge or non-targeted CT knowledge could interfere with students' ability to respond to an item) and provided evidence that specific CT assessment items seemed to elicit the relevant KSAs in the specific LT.

We have revised the assessments following our 2018–2019 pilot implementation, including editing individual assessment items, the overall instruments, and the design patterns. Such iterative design is critical in developing and validating assessments. We are currently analyzing data collected from the use of these revised assessments during AY 2019–2020. The analytic and empiric approach we describe in this paper can be used for the other items with these new data, including items that appeared on the other four assessment instruments (i.e., the Mid and Late instruments in Grades 3 and 4).

Because this 2019–2020 data collection involved a greater number of teachers and students, we are hopeful that these larger sample sizes will enable more sophisticated psychometric modeling, potentially including structural analysis (e.g., to determine dimensionality and investigate the use of sub-scores), further IRT analysis (using 2PL models), and diagnostic classification modelling (Pellegrino et al., 2014). Such data would allow us to advance and support additional claims about the appropriateness of the instruments (i.e., the specific assemblages of items), the utility and appropriateness of calculating sub-scores tied to individual LTs, and the use of these overall and/or sub-scores to understand student performance and growth over time. However, we also know that further validation efforts must be carefully considered in light of the claims being made for the assessments and uses. As Douglas and Purzer (2015) point out, it can be tempting to equate the indiscriminate application of psychometric approaches with validation, but doing so is a mistake. Our design of the CT items and instruments allows for various psychometric analyses (e.g., factor analysis, 2PL IRT, diagnostic classification modeling) that could be informative and support interesting claims. However, expectations for engaging in these efforts must be tempered with a dose of reality. The limiting factor is the number of student responses needed to perform these analyses. For data to be meaningful, students must have had the appropriate opportunities to learn. Therefore, data collection is limited to teachers and students that are using the curricular lessons, and faces the realities of limited time and student attention. Elementary students will not sit for long assessments, and teachers are reluctant to give up instructional time for assessments that do not translate into grades for students. These are real and valid constraints on data collection and limit the potential for using sophisticated psychometric approaches (more common in large scale testing) with data collected from classroom assessments completed as part of students' in-class activities.

In addition to our efforts to further validate the assessments, we are also interested in exploring different uses of the assessments, particularly in ways that can support teachers' instruction and students' learning. Because we used a design method that entails documenting the rationale underlying design decisions and item alignment, these assessments do not have to be used as "intact" instruments. Instead they could be disassembled and reassembled based on the specific lessons that are taught or the specific content that one wants to assess. For instance, one could assemble items on the basis of the LT from which they were developed in order to create a progress variable against which students' performance could be measured and tracked (Pellegrino et al., 2016). In the future, these assessments might be used by teachers as diagnostic instruments that could be used formatively, summatively, and/or in a pre–post design. Such uses are enabled by the design process but would also require additional validation. To date we have focused on using a principled design process to develop the assessment items and associated scoring criteria/rubrics, and on collecting initial evidence to support the claim that the CT assessment items measure students' performance with respect to the LTs (topics) that students were introduced to during the lessons. How to support teachers in using these assessments formatively represents future work.

Final thoughts

As the field seeks to make advances in CT assessment, attention must be paid to the connections between learning and assessment, especially as they are mediated by processes of curriculum design and instruction. As argued in *Knowing What Students Know: The Science and Design of Educational Assessment* (Pellegrino et al., 2001), we need to strive for coordination among curriculum, instruction, and assessment. Such coordination is only possible when all three are guided by theories, models, and data about learning and knowing in the domain of interest. We have adhered to this conceptualization in our initial work on the design of CT assessments. Our design process allowed us to create assessment items and organize them into multiple instruments, mapped to different phases of instruction, where the content assessed was based on the same learning goals (LTs) that were used to design the curricular lessons. Although some items were identified for revision, the majority performed well. The process of using the LTs to develop assessments tied to curricular lessons is one of several possible approaches to the assessment of CT. Rather than trying to provide a general evaluation of students' CT, our assessments were designed to cover multiple facets of CT (i.e., the different LTs) that students would be expected to learn through instruction based on a set of carefully designed lessons. Furthermore, the assessments were not developed directly from the content of lessons but focused on the CT constructs embedded in the lessons. As such, they are best considered proximal assessments (Ruiz-Primo et al., 2002), and are appropriate for measuring the effect of the curriculum and instruction while also serving as a measure of near transfer (Pellegrino et al., 2014).

Note

1. A design pattern for the debugging LT was not developed due to concerns about the difficulty of attempting to assess debugging, especially when using paper-and-pencil assessments.

Acknowledgments

We would also like to acknowledge the vital assistance of Carla Strickland, Diana Franklin, Katie Rich, & Susan Krause in reviewing assessments and design patterns, Donna Eatinger, Todd Lash, & Gakyung Jeong in facilitating data collection, Ruohan Liu in scoring student responses, and Henrique Cirne-Lima in preparing Figure 1. A previous version of this paper presenting preliminary design and empirical results appeared in the ICLS 2020 conference proceedings.

Disclosure statement

No potential conflict of interest was reported by the authors.

Funding

This work was supported by the National Science Foundation under grant number 1932920. Any opinions, findings, and conclusions or recommendations expressed in this material are those of the author(s) and do not necessarily reflect the views of the National Science Foundation.

ORCID

Brian D. Gane http://orcid.org/0000-0002-4454-569X
Maya Israel http://orcid.org/0000-0003-0302-6559
Feiya Luo http://orcid.org/0000-0002-3037-085X

References

AERA, APA, & NCME. (2014). *Standards for educational and psychological testing (2014 ed.)*. American Educational Research Association.

Baker, F. B. (2001). The basics of item response theory (2nd edition). *ERIC Clearinghouse on Assessment and Evaluation*. http://files.eric.ed.gov/fulltext/ED458219.pdf

Barr, D., Harrison, J., & Conery, L. (2011). Computational thinking: A digital age skill for everyone. *Learning & Leading with Technology, 38*(6), 20–23. http://files.eric.ed.gov/fulltext/EJ918910.pdf

Barr, V., & Stephenson, C. (2011). Bringing computational thinking to K-12: What is Involved and what is the role of the computer science education community? *ACM Inroads*, *2*(1), 48–54.https://doi.org/10.1145/1929887.1929905

Bers, M. U., Flannery, L., Kazakoff, E. R., & Sullivan, A. (2014). Computational thinking and tinkering: exploration of an early childhood robotics curriculum. *Computers & Education*, *72*, 145–157. https://doi.org/10.1016/j.compedu.2013.10.020

Bienkowski, M., Snow, E., Rutstein, D., & Grover, S. (2015). *Assessment design patterns for computational thinking practices in secondary computer science: A first look*. (SRI technical report). Menlo Park, CA, USA: SRI International.http://pact.sri.com/resources.html

Boyer, T. W., & Levine, S. C. (2012). Child proportional scaling: is 1/3= 2/6= 3/9= 4/12? *Journal of Experimental Child Psychology*, *111*(3), 516–533. https://doi.org/10.1016/j.jecp.2011.11.001

Brennan, K., & Resnick, M. (2012). New frameworks for studying and assessing the development of computational thinking. Paper presented at the *2012 Annual Meeting of the American Educational Research Association*, Vancouver, Canada.

Clements, D. H., Sarama, J., Baroody, A. J., & Joswick, C. (2020). Efficacy of a learning trajectory approach compared to a teach-to-target approach for addition and subtraction. *ZDM – Mathematics Education*, *52*(4), 637–648. https://doi.org/10.1007/s11858-019-01122-z

Computer Science Teachers Association [CSTA]. (2017). *CSTA K-12 computer science standards, revised 2017*. Computer Science Teachers Association, USA.https://www.csteachers.org/page/standards

Crocker, L. (2006). Introduction to measurement theory. In J. Green, L. G. Camilli, & P. B. Elmore (Eds.), *Handbook of complementary methods in education research* pp.(371–384). Erlbaum.

Denny, P., Luxton-Reilly, A., & Simon, B. (2008). Evaluating a new exam question: Parsons problems. In *Proceedings of the fourth international workshop on computing education research* pp.(113–124) https://doi.org/10.1145/1404520.1404532

Douglas K., A., Gane, B. D., Neumann, K., & Pellegrino, J. W., (2020). Contemporary methods of assessing integrated STEM competencies. In C. C. Johnson, M. Mohr-Schroeder, T. Moore, L. Bryan, & L. English (Eds.) *Handbook of research on STEM education* pp.(234–254). Routledge/Taylor & Francis.https://doi.org/10.4324/9780429021381

Douglas, K. A., & Purzer, Ş. (2015). Validity: meaning and relevancy in assessment for engineering education research. *Journal of Engineering Education*, *104*(2), 108–118. https://doi.org/10.1002/jee.20070

Embretson, S. E., & Reise, S. P. (2000). Item response theory for psychologists. Erlbaum.

English, L. D. (2017). Advancing elementary and middle school STEM education. *International Journal of Science and Mathematics Education*, *15*(S1), 5–24. https://doi.org/10.1007/s10763-017-9802-x

Ericson, B. J., Margulieux, L. E., & Rick, J. (2017). Solving parsons problems versus fixing and writing code. In *Proceedings of the 17th koli calling international conference on computing education research* pp.(20–29), Koli, Finland.

Gane, B. D., Zaidi, S. Z., & Pellegrino, J. W. (2018). Measuring what matters: Using technology to assess multidimensional learning. *European Journal of Education*, *53*(2), 176–187. https://doi.org/10.1111/ejed.12269

Harwell, M., Moreno, M., Phillips, A., Guzey, S. S., Moore, T. J., & Roehrig, G. H. (2015). A study of STEM assessments in engineering, science, and mathematics for elementary and middle school students. *School Science and Mathematics*, *115*(2), 66–74. https://doi.org/10.1111/ssm.12105

Israel, M., Pearson, J. N., Tapia, T., Wherfel, Q. M., & Reese, G. (2015). Supporting all learners in school-wide computational thinking: A cross-case qualitative analysis. *Computers & Education*, *82*, 263–279. https://doi.org/10.1016/j.compedu.2014.11.022

K-12 Computer Science Framework (2016). *K-12 computer science framework*.https://www.k12cs.org/

Kane, M. T. (1992). An argument-based approach to validation. *Psychological Bulletin*, *112*, 527–535. https://doi.org/10.1037/0033-2909.112.3.527

Kane, M. T. (2006). Validation. In R. L. Brennan (Ed.) *Educational measurement* (4th Ed., pp.17–64). Praeger Publishers.

Lee, I., Martin, F., & Apone, K. (2014). Integrating computational thinking across the K–8 curriculum. *ACM Inroads*, *5*(4), 64–71. https://doi.org/10.1145/2684721.2684736
Luo, F. (2020). *Exploring elementary students' computational thinking leveraging affordances of learning trajectories* [Unpublished doctoral dissertation]. University of Florida.
Luo, F., Israel, M., Liu, R., Yan, W., Gane, B., & Hampton, J. (2020). Understanding students' computational thinking through cognitive interviews: A learning trajectory-based analysis. In *Proceedings of the 51st ACM technical symposium on computer science education pp.* (919–925). Portland, OR, USA: Association for Computing Machinery. https://doi.org/10.1145/3328778.3366845
Mair, P., & Hatzinger, R. (2007). Extended rasch modeling: the eRm package for the application of IRT models in R. *Journal of Statistical Software, 20*(9), 1–20. http://dx.doi.org/10.18637/jss.v020.i09
Maloney, A. P., Confrey, J., & Nguyen, K. H.(Eds.). (2014). *Learning over time: Learning trajectories in mathematics education*. Information Age Publishing.
Maloney, J., Resnick, M., Rusk, N., Silverman, B., & Eastmond, E. (2010). The scratch programming language and environment. *ACM Transactions on Computing Education*, *10*(4), 1–15. https://doi.org/10.1145/1868358.1868363
Mislevy, R. J., & Haertel, G. D. (2006). Implications of evidence-centered design for educational testing. *Educational Measurement: Issues and Practice*, *25*(4), 6–20. https://doi.org/10.1111/j.1745-3992.2006.00075.x
Mislevy, R. J., Steinberg, L. S., & Almond, R. G. (2003). On the structure of educational assessments. *Measurement: Interdisciplinary Research & Perspective*, *1*(1), 3–62. https://doi.org/10.1207/S15366359MEA0101_02
Moreno-León, J., Robles, G., & Román-González, M. (2015). Dr. scratch: Automatic analysis of scratch projects to assess and foster computational thinking. *RED. Revista De Educación a Distancia*,(46), 1–23. https://revistas.um.es/red/article/view/240251
Pellegrino, J. W., DiBello, L. V., & Brophy, S. P. (2014). The science and design of assessment in engineering education. In A. Johri & B. M. Olds (Eds.), *Cambridge handbook of engineering education research* pp.(571–598). Cambridge University Press.
Pellegrino, J. W., Chudowsky, N., & Glaser, R. (Eds.). (2001). *Knowing what students know: The science and design of educational assessment*. The National Academies Press.
Pellegrino, J. W., DiBello, L. V., & Goldman, S. R. (2016). A framework for conceptualizing and evaluating the validity of instructionally relevant assessments. *Educational Psychologist*, *51*(1), 59–81. https://doi.org/10.1080/00461520.2016.1145550
Price, T. W., & Barnes, T. (2015). Comparing textual and block interfaces in a novice programming environment. In *Proceedings of the eleventh annual international conference on international computing education research* pp.(91–99). Omaha, USA.
Resnick, M., Maloney, J., Monroy-Hernández, A., Rusk, N., Eastmond, E., Brennan, K., … Kafai, Y. (2009). Scratch: Programming for all. *Communications of the ACM*, *52*(11), 60–67. https://doi.org/10.1145/1592761.1592779
Rich, K. M., Franklin, D., Strickland, C., Isaacs, A., & Eatinger, D. (2020). *A learning trajectory for variables Based in computational thinking literature: Using levels of thinking to develop instruction, Computer Science Education. https://doi.org/10.1080/08993408.2020.1866938.*
Rich, K. M., Binkowski, T. A., Strickland, C., & Franklin, D. (2018). Decomposition: A k-8 computational thinking learning trajectory. In *Proceedings of the 2018 ACM conference on international computing education research* pp.(124–132). Espoo, Finland.
Rich, K. M., Strickland, C., Binkowski, T. A., & Franklin, D. (2019). A k-8 debugging learning trajectory derived from research literature. In *Proceedings of the 50th ACM technical symposium on computer science education* pp.(745–751). Minneapolis, MN.
Rich, K. M., Strickland, C., Binkowski, T. A., Moran, C., & Franklin, D., (2017). K-8 learning trajectories derived from research literature: Sequence, repetition, conditionals. In *Proceedings of the 2017 ACM conference on international computing education research* (pp.182–190). Tacoma, WA, USA: ACM. http://dx.doi.org/10.1145/3105726.3106166
Román-González, M., Moreno-León, J., & Robles, G. (2019). Combining assessment tools for a comprehensive evaluation of computational thinking interventions. In S.-C. Kong & H.

Abelson (Eds.) *Computational thinking education* pp.(79–98). Springer, Singapore. https://doi.org/10.1007/978-981-13-6528-7_6

Román-González, M., Pérez-González, J. C., & Jiménez-Fernández, C. (2017). Which cognitive abilities underlie computational thinking? Criterion validity of the computational thinking test. *Computers in Human Behavior, 72*, 678–691. https://doi.org/10.1016/j.chb.2016.08.047

Ruiz-Primo, M. A., Shavelson, R. J., Hamilton, L., & Klein, S. (2002). On the evaluation of systemic science education reform: Searching for instructional sensitivity. *Journal of Research in Science Teaching, 39*(5), 369–393. https://doi.org/10.1002/tea.10027

Santo, R., DeLyser, L. A., Ahn, J., Pellicone, A., Aguiar, J., & Wortel-London, S. (2019). Equity in the who, how and what of computer science education: K12 school district conceptualizations of equity in 'CS for all' initiatives. In *2019 research on equity and sustained participation in engineering, computing, and technology (RESPECT)* pp.(1–8). Minneapolis, MN, USA: IEEE.

Seiter, L., & Foreman, B. (2013). Modeling the learning progressions of computational thinking of primary grade students. In *Proceedings of the ninth annual international ACM conference on international computing education research* pp.(59–66). San Diego, CA, USA. https://doi.org/10.1145/2493394.2493403

Shute, V. J., Sun, C., & Asbell-Clarke, J. (2017). Demystifying computational thinking. *Educational Research Review, 22*, 142–158. https://doi.org/10.1016/j.edurev.2017.09.003

Snow, E., Rutstein, D., Basu, S., Bienkowski, M., & Everson, H. T. (2019). Leveraging evidence-centered design to develop assessments of computational thinking practices. *International Journal of Testing, 19*(2), 103–127. https://doi.org/10.1080/15305058.2018.1543311

Streveler, R. A., Miller, R. L., Santiago-Roman, A. I., Nelson, M. A., Geist, M. R., & Olds, B. M. (2011). Rigorous methodology for concept inventory development: using the 'assessment triangle' to develop and test the thermal and transport science concept inventory (TTCI). *International Journal of Engineering Education, 27*(5), 1-17. https://www.ijee.ie/contents/c270511.html

Strickland, C., Rich, K. M., Eatinger, D., Lash, T., Isaacs, A., Israel, M., & Franklin, D. (2021). Action fractions: The design and pilot of an integrated math+CS elementary curriculum based on learning trajectories. In *Proceedings of the 2021 ACM SIGCSE technical symposium on computer science education*. ACM.

Tang, X., Yin, Y., Lin, Q., Hadad, R., & Zhai, X. (2020). Assessing computational thinking: A systematic review of empirical studies. *Computers & Education, 148*, 103798. https://doi.org/10.1016/j.compedu.2019.103798

Viera, A. J., & Garrett, J. M. (2005). Understanding interobserver agreement: the kappa statistic. *Family Medicine, 37*(5), 360–363. https://www.stfm.org/familymedicine/vol37issue5/Viera360 , https://fammedarchives.blob.core.windows.net/imagesandpdfs/pdfs/FamilyMedicineVol37Issue5Viera360.pdf

Voogt, J., Fisser, P., Good, J., Mishra, P., & Yadav, A. (2015). Computational thinking in compulsory education: towards an agenda for research and practice. *Education and Information Technologies, 20*(4), 715–728. https://doi.org/10.1007/s10639-015-9412-6

Voskoglou, M. G., & Buckley, S. (2012). Problem solving and computational thinking in a learning environment. *Egyptian Computer Science Journal, 36*(4), 28–46. https://arxiv.org/abs/1212.0750

Weintrop, D., Beheshti, E., Horn, M., Orton, K., Jona, K., Trouille, L., & Wilensky, U. (2016). Defining computational thinking for mathematics and science classrooms. *Journal of Science Education and Technology, 25*(1), 127–147. https://doi.org/10.1007/s10956-015-9581-5

Wing, J. M. (2006). Computational thinking. *Communications of the ACM, 49*(3), 33–35. https://doi.org/10.1145/1118178.1118215

Wing, J. M. (2014). Computational thinking benefits society. *40th Anniversary Blog of Social Issues in Computing*, 2014. http://socialissues.cs.toronto.edu/index.html%3Fp=279.html

Yadav, A., Hong, H., & Stephenson, C. (2016). Computational thinking for all: Pedagogical approaches to embedding 21st century problem solving in K-12 classrooms. *TechTrends, 60*(6), 565–568. https://doi.org/10.1007/s11528-016-0087-7

Zhang, L., & Nouri, J. (2019). A systematic review of learning computational thinking through Scratch in K-9. *Computers & Education, 141*, 103607 https://doi.org/10.1016/j.compedu.2019.103607

A principled approach to designing computational thinking concepts and practices assessments for upper elementary grades

Satabdi Basu, Daisy W. Rutstein, Yuning Xu, Haiwen Wang and Linda Shear

ABSTRACT

Background and Context: In today's increasingly digital world, it is critical that all students learn to think computationally from an early age. Assessments of Computational Thinking (CT) are essential for capturing information about student learning and challenges. When programming is used as a vehicle to foster CT skills, assessment of CT skills needs to measure understanding of programming concepts alongside general problem-solving practices applicable in contexts beyond programming.

Objective: In this paper, we discuss the development of and results from separate validated assessments of programming concepts and generic CT practices for 4th-6th grade students.

Method: An Evidence-Centered Design approach was used for the development and validation of the assessments, that were used with 15,000 upper elementary students in Hong Kong.

Findings: Results from assessment implementation indicate that the assessments worked as designed and reveal student challenges with CT concepts and practices.

Implications: These results point to the utility of our assessment as a curricular tool and the need for emphasizing certain CT concepts and practices in future curricular initiatives and teacher professional development.

1 Introduction

In today's increasingly digital world driven by computation, it is critical that all students learn to recognize and leverage the power of computation from an early age in order to better understand and solve real-world problems. Computational Thinking (CT) focuses on the ability to problem-solve in various disciplines by using computational skills that may or may not include programming (Aho, 2012; Wing, 2006). CT involves leveraging computational affordances, formulating problems such that they can be solved computationally, and identifying which problems are best solved by a computational agent versus a human being. Initiatives to equip all K-12 students with CT skills have gained momentum worldwide in the last decade (Bocconi et al., 2016).

With the increase in emphasis on CT skills in formal K-12 education (K-12 Computer Science Framework, 2016; NGSS Lead States, 2013) comes an increase in the need for measures that support valid inferences about students' abilities related to these skills. Developing valid and reliable measures for assessing students' CT skills helps evaluate the effectiveness of different CT curricula and instructional approaches, and supports teachers in better understanding their students' progress on and difficulties with specific computational concepts and practices. Assessment of CT skills needs to focus on students' abilities to engage in problem-solving practices such as logical thinking, formulation of a problem as a set of computational steps, pattern recognition, abstraction and generalization, decomposition and modularization, data collection and organization; data-based decision making, and systematic incremental testing and debugging (Atmatzidou & Demetriadis, 2016; Barr et al., 2011; Csizmadia et al., 2015).

While CT skills are not equivalent to programming skills, programming does offer an important context to promote CT in K-12 settings. Applying CT skills in the context of programming involves the ability to understand programming representations and use programming constructs, also known as CT concepts (Brennan & Resnick, 2012). When programming is used as a vehicle to foster CT skills, assessment of CT skills needs to include assessment of understanding of programming or CT concepts alongside assessment of general problem-solving practices such as reusing and remixing, testing and debugging, and abstracting and modularizing that are important in contexts beyond programming. These practices reflect skills that students should be able to do, show, engage in, and explain their thinking about across diverse domains and contexts (not just programming contexts) and are commonly known as CT practices (Brennan & Resnick, 2012).

Integrated assessment of CT concepts and practices is often ideal and mirrors the CT process more authentically, but there is also benefit to separate assessments for CT concepts and practices. For example, assessing CT practices separately in non-programming contexts helps focus on students' transferable problem-solving skills outside the programming context and enables assessment without collecting and analysing students' process data during programming. In this paper, we discuss the use of a principled assessment design framework for the design and development of separate validated CT practices and CT concepts assessments for 4th-6th grade students who are learning CT skills in a programming context. The assessments were developed as part of a larger evaluation effort for CoolThink@JC – a 4-year pilot program to teach programming, foster problem-solving skills, and generate interest in computing for students in grades 4–6 (Primary 4–6) in Hong Kong schools. We describe detailed assessment targets for various CT practices and concepts, present sample assessment tasks, and discuss results from one implementation of the assessments, revealing some common student challenges. Alongside contributing to the literature on CT assessments for early grades, our observations from the assessment implementations contribute to the literature on student understanding of CT, and point to the need for emphasizing certain aspects of CT practices and concepts in curricula and teacher preparation.

2 Background

2.1 Existing work on CT assessments

There is a growing body of literature around K-12 assessments for CT which involves wide variation in aspects of CT emphasized, the scale of application of the assessments, and the level of psychometric rigor established in the assessments. Currently, most K-12 CT assessments are computer-based, non-automatically scored, and measure cognitive rather than interpersonal or intrapersonal CT skills (Cutumisu et al., 2019). Algorithms, logical thinking, control (loops, conditionals), data (variables) and modularity are the most commonly assessed constructs in CT assessments, while block-based languages and generic representations not tied to specific programming languages are the most commonly used representations in K-12 CT assessments (Cutumisu et al., 2019). Selected response items, programming projects, surveys and self-report scales represent the most common format used for CT assessment instruments (Alves et al., 2019; Cutumisu et al., 2019; Martins-Pacheco et al., 2019). When CT is fostered through programming contexts, assessment of CT typically focuses on students' understanding and ability to use programming constructs (CT concepts) and less frequently focuses on assessing students' general problem-solving practices in non-programming contexts (Basu et al., 2020). When developing CT assessments, it is important to be mindful of the context (programming, data science, tangible computing, etc.) in which students engage with CT since the context can influence the target CT skills and cognitive abilities that need to be measured (Knee et al., 2006).

2.1.1. Existing assessments for CT concepts

Several CT assessments focus primarily on students' understanding of programming constructs such as sequences, loops, variables and conditionals (e.g., Grover & Basu, 2017; Román-González et al., 2016; Weintrop & Wilensky, 2015). Many of these assessments are tied to specific programming languages like Scratch, App Inventor, Alice, Python, etc. while many others use generic representations not tied to specific programming languages. These efforts use different formats like selected response, constructed response, and Parsons problems, and assess students not only on knowledge of programming concepts but also the ability to use the concepts to comprehend code, debug code, and complete or create code (Zur-Bargury et al., 2013). The CTt assessment (Román-González et al., 2016) is a well-known and validated 28 item CT concept instrument that uses block-based programming-like elements. It is designed to assess students between 12 and 14 years of age (middle school), and assesses the following seven computational concepts: basic directions and sequences, loops-repeat times, loops-repeat until, if-simple conditional, if/else-complex conditional, while conditional, and simple functions.

Some other efforts in this space use performance assessments based on the analysis of code developed by students or search for presence of specific programming constructs in students' programming artifacts (Seiter & Foreman, 2013). The presence or absence of programming constructs can provide inaccurate evidence of what a student knows (Dorn et al., 2007), and provides no evidence for the computational processes that students employ to reach the final product. Performance assessments involve analysis of students' code against certain criteria and can help infer students' computational competencies (Moreno-León et al., 2017), but assessing complex, ill-structured problems for which no single correct solution exists tends to be challenging (Funke, 2012). Tools such as *Dr. Scratch* (Moreno-León et al., 2015, 2016) offer automatic online analysis of students' Scratch programming artifacts based on different dimensions of CT competency: abstraction and problem decomposition, parallelism, logical thinking, synchronization, flow control, user interactivity, and data representation.

2.1.2. Assessing CT practices in programming contexts

CT practices constitute a high but important bar to reach for in CT assessments (Snow et al., 2017; Yadav et al., 2015). They also reflect a broader shift in other fields (science and engineering) away from measurements of factual knowledge and toward knowledge and skill application for problem-solving. While CT concepts assessments often measure concept knowledge as well as ability to apply the knowledge, the process that leads students to their final artifacts also provides important information. Some work has been done in assessing students' CT practices by observing students (in real-time or via video recordings) or conducting artifact-based interviews (Brennan & Resnick, 2012) where students are asked to explain a given artifact, extend it, fix it, and add a new feature. Such assessment methods are time-consuming and not scalable. Another body of work has focused on analyzing process data logged as students build computational solutions using a combination of learning analytics and educational data mining techniques (Basu et al., 2017; Berland et al., 2013; Blikstein et al., 2014; Grover et al., 2017; Koh et al., 2010; Werner et al., 2012). Some of these efforts place students in scenarios carefully designed to elicit desired CT practices. For example, the Fairy Assessment (Werner et al., 2012) presented middle school learners with partially complete or buggy Alice programs that they had to fix and complete, eliciting the practices of code comprehension, algorithmic thinking, and testing and debugging. However, these assessment methods are not suitable for use as a standard test, especially as a pre-test measure before students are introduced to computational tools, and involve technological and logistical requirements and data privacy issues.

2.1.3. Assessing CT practices in non-programming contexts

Another strand of assessment development has focused on developing CT Practices instruments that are multi-lingual, can be used across contexts and curricula, and can be used as both pre and post measures. However, few validated assessments of CT practices are known in this strand, and they are particularly sparse for early grades. PACT is a well-known assessment specifically designed for the high school Exploring Computer Science (ECS) curriculum (Bienkowski et al., 2015; Snow et al., 2017). The PACT assessments for ECS Unit 2: Problem Solving are paper/pencil tasks in this assessment strand that ask students to engage with generic CT practices, while providing supports for student's knowledge of CT concepts. Bebras items (Dagienė & Futschek, 2008) are also widely used as assessments of CT Practices, though developed as part of a competition. The items do not depend on prior coding knowledge and use everyday visual metaphors in their puzzle-like problem-solving tasks, but do not always have a 1–1 mapping to CT practices. Recently, Wiebe et al. (2019) combined some Bebras items with some CTt items (Román-González et al., 2016) to create a "lean" CT assessment for middle school. Another example of a CT practices assessment in this strand can be found in Chen et al.'s (2017) pre-post CT assessment for a humanoid robotics curriculum. The 23-item assessment measures fifth-grade students' CT skills contextualized in two types of CT applications: coding in robotics (CT concepts) and reasoning of everyday events (CT practices in non-programming contexts).

In this paper, we describe the design and development of separate CT practices and CT (programming) concepts assessments for upper-elementary students, both using computer-based, automatically scorable formats. Our CT concepts assessment measures programming constructs using Scratch and App Inventor while the CT Practices assessment tasks are not tied to any programming representation, block-based or otherwise.

2.2 Evidence-centered design

A principled assessment design framework, Evidence-Centered Design (ECD), (Mislevy & Haertel, 2006), was used to develop the CT (programming) concepts and CT practices assessment tasks. ECD promotes coherence in the design of assessment tasks and rubrics and the interpretation of students' performances by explicitly linking claims about student learning, evidence from student work products, and design features of tasks that elicit the desired evidence. It is principled in that there are explicit steps that focus the developer on critical aspects of the assessment that must be addressed through the design process. ECD begins with a domain analysis, which entails gathering and organizing information on the domain to be assessed (CT being the domain in this case). The focus is on the identification of key aspects of the domain, the relationship between these aspects, and the representation of these aspects. This is followed by domain modeling where details of the assessment argument are specified. These details include articulating (a) the student model (defining specific learning targets and their relationships to each other), (b) the evidence model (identifying what students should produce and how these products should be scored to reflect on the learning targets), and (c) the task model (specifying characteristics of tasks that would allow students to produce the desired evidence). This articulation is at a general enough level that multiple tasks can be developed based on the specifications, but specific enough to highlight the connection among the learning targets, the evidence needed from the tasks, and the task design specifications. The information specified here informs the development of tasks and rubrics. The use of an ECD process for assessment development focuses the designer on the critical aspects of the assessment and helps ensure that the assessment matches its intended purpose or goal. This approach supports the validity of the assessment (Mislevy, 2007), by providing evidence of the construct validity of the assessment. ECD has been used to develop assessments for different disciplines including Computer Science (Basu et al., 2020; Bienkoski et al., 2015; Snow et al., 2019), for example, assessments for the ECS curriculum (Goode et al., 2012).

3 Methodology

In this section, we discuss how an ECD process was applied to design and validate assessments for CT practices and CT concepts and address the following research questions:

1. How can we design a valid assessment instrument for CT practices such as algorithmic thinking, abstraction, testing and debugging, and reusing and remixing, that can be used at scale for 4th – 6th grade students?

2. How can we design a valid assessment instrument for CT (programming) concepts such as sequences, data structures, conditionals, iterations, and procedures?

3. What do these assessment tasks tell us about student understanding and challenges related to the use of CT concepts and CT practices?

3.1 The CoolThink@JC pilot initiative

The development of the CT practices and concepts assessments was part of a larger 4-year pilot initiative – the CoolThink@JC initiative – to teach computer programming and computational thinking to students aged 9–11 (students in primary 4, 5, and 6) in 32 Hong Kong schools. The CoolThink@JC pilot initiative is designed to achieve the following outcomes:

- Increase students' programming content knowledge related to computational thinking (CT concepts),
- Improve student's problem-solving/logical thinking skills (CT practices), and
- Generate interest and motivation for computational thinking (CT perspectives).

The pilot lessons consist of 3 levels, one each for primary 4, 5 and 6 (Kong et al., 2017). Each level builds on the previous one, adding concepts of increasing complexity over time. Level 1 lessons primarily use the Scratch visual programming language while Level 2 and Level 3 lessons mainly use the App Inventor platform. The initiative was rolled out over several years, with the first cohort of schools starting one year before the second cohort of schools. Students in the first cohort of schools started in primary grade 4 and continued for three years. Students in the second cohort of schools also started in primary grade 4 and had gone through two years of the program at the time of the analysis presented in this paper.

Several instruments were developed as part of the evaluation of this initiative. In this paper, we limit our discussion to the design of an assessment for CT concepts and an assessment for CT practices. While CT practices are often assessed along with CT concepts, one goal of the evaluation was to determine if students improved in their engagement with generic CT practices that could potentially be applicable in contexts beyond programming. Therefore, the CT practices assessment was designed as a separate assessment that did not rely on programming structures. Since both the CT concepts and practices assessment instruments were developed to evaluate the impact of the CoolThink@JC initiative, they were designed to align with the learning goals of the initiative and the curriculum used as part of the initiative. While the choice of which aspects of CT to focus on in the assessments and their categorization was guided by the learning goals of the initiative, we found that the target CT constructs were ones commonly highlighted in a number of CT interventions (Basu et al., 2017; Chen et al., 2017; Leonard et al., 2016; Mouza et al., 2016; Román-González et al., 2016; Strawhacker et al., 2018).

The assessments were developed in the first two years of the initiative and were piloted in Fall 2017. The assessments were then administered in Spring of 2018 and again in Spring of 2019. With around 15,000 students participating in the initiative in any given year, it was critical that the assessments could be administered on computers and the scoring of the assessments was quick and reliable. Hence, we opted for selected response formats for most questions, and slightly different, but easily scorable formats for other questions (e.g., reordering given actions, dragging and dropping blocks into correct categories). Further information on assessment development for each of the instruments is described in the following sections. The results presented for both assessment instruments are from the Spring 2019 administration.

3.2 Designing a CT Practices assessment instrument

When developing the CT Practices assessment using an ECD approach, the first step was to identify and obtain a deep understanding of the practices that were to be included. Four categories of CT practices (CTP) were chosen as the focus of the CoolThink@JC initiative and therefore as the focus of our assessment (Basu et al., 2020). Three of the practices were chosen based on the framework described by Brennan and Resnick (2012) – *Reusing and Remixing, Abstraction and Modularization,* and *Testing and Debugging*. Another practice described in the Brennan and Resnick (2012) framework – *Being incremental and iterative* – was not selected as a target for our assessment since it is best assessed over time in a dynamic context instead of through a one-time static assessment. A fourth practice, *Algorithmic Thinking*, was derived from commonly used skills used to characterize CT (Cutumisu et al., 2019). The four selected categories of practices are not entirely mutually independent but are distinct enough to be assessed separately.

We defined a scope for each of the four practices based on a domain analysis that focused on how these practices were described in (a) the Brennan and Resnick (2012) paper, (b) the CoolThink@JC curriculum, (c) the K-12 CS framework (K-12 Computer Science Framework, 2016), and (d) existing design patterns (e.g., design pattern for algorithmic thinking – Bienkowski et al., 2015). Next, we elaborated a set of focal knowledge, skills and abilities (FKSAs) that would be expected of students engaging with each of the practices. A subset of the generated FKSAs was selected for the assessment tasks. FKSAs that were not critical for the CoolThink@JC initiative were set aside. Next, the FKSAs were evaluated in relation to the assessment format constraints described in Section 3.1. Our final set of 18 FKSAs are marked with a *.

CTP1 – **Reusing and remixing** refers to the practice of making something by building on or adapting existing projects or ideas, either your own, or from other people. It involves understanding programs or problem solutions possibly created by others, and identifying reusable parts, instead of building solutions from scratch. This practice involves ethical concerns related to appropriately acknowledging the author of the solution being reused or remixed. The FKSAs are:

(1) Ability to evaluate existing programs and problem solutions in light of a new problem or purpose. *
(2) Ability to reuse (without modifications) all or part of existing solutions in new solutions.*
(3) Ability to adapt all or part of existing solutions to be applicable for new solutions.*
(4) Ability to evaluate tradeoffs between reusing and/or remixing and creating a problem solution anew. *

(5) Ability to describe how the cultural context and audience for which computing artifacts are developed impacts their design and re-use.
(6) Ability to evaluate and address legal and ethical concerns raised by reusing others' solutions.

CTP2 – **Algorithmic Thinking** is a method for problem solving or addressing a need by clearly defining the steps needed to reach the problem solution. This includes students being able to articulate the initial conditions, the goal state, and instructions that would take them from the initial condition to the goal state. It also involves being able to compare different approaches to solving a problem based on various criteria. The FKSAs are:

(1) Recognize appropriate and worthwhile opportunities to apply computation to solve a problem.
(2) Ability to describe the differences between how a computer and a human perform a task, including the idea that computers follow instructions that must be precise and not open to interpretation.
(3) Ability to recognize relevant information from a problem to identify inputs and goals of an algorithm, boundary conditions, stopping conditions, and constraints such as cost, time or delivery platform.*
(4) Ability to summarize the behavior of an algorithm when given a specific set of inputs.*
(5) Ability to decompose a problem meaningfully and evaluate each part to determine whether a computational solution is appropriate.
(6) Ability to design a problem solution that handles the desired range of inputs and is able to deal with boundary conditions/edge cases.*
(7) Ability to generate multiple ways to solve a problem.
(8) Ability to compare the trade-offs between multiple approaches/techniques to implementing a problem based on certain evaluation criteria.*

CTP3 – **Abstraction and Modularization** is a means of hiding complexity and building something large by putting together collections of smaller parts. Abstractions are formed by identifying patterns and extracting common features from specific examples to create generalizations. The FKSAs are:

(1) Ability to describe how a computer model or simulation abstracts real world phenomena.
(2) Ability to explain how computers represent objects as data (e.g., media files, bar codes).
(3) Ability to identify or describe patterns. *
(4) Ability to break down a problem with the goal of collaboratively building a solution with a team.*
(5) Ability to describe functions, classes, and methods to divide a complex problem into simpler parts.
(6) Ability to compose a solution by combining subparts that will lead to a solution to the problem as a whole.*

(7) Ability to reason about a problem at multiple levels of detail, including describing components at each level and relationships among components and levels.
(8) Ability to explain how parameterization can be used to generalize a specific solution.
(9) Ability to design an abstraction and evaluate the appropriateness of an abstraction for a given purpose based on the assumptions and simplifications made.*

CTP4 – **Testing and Debugging**. Testing involves ensuring that a solution works by checking it against its intended outcome with a systematic set or range of inputs. Debugging is a cyclic activity involving testing, inspection of the correctness of the output, determining what caused any perceived errors, and correcting the problem solution and retesting until the output matches the desired outcome. The FKSAs are:

(1) Knowledge that debugging is an incremental and iterative process.*
(2) Ability to identify erroneous behavior or problem requirements not addressed by a solution.*
(3) Ability to vary parameters and determine conditions under which the solution works and does not work.*
(4) Ability to evaluate a problem solution in parts, making sure each part works individually, before evaluating interactions between the parts.
(5) Ability to vary one variable at a time to evaluate its effects on the problem solution.
(6) Ability to describe a systematic method for error-detection (test cases, unit testing). *
(7) Ability to manually trace code, inspect variable values and make use of available debugging technology.
(8) Ability to map pieces of the problem solution to resultant behavior and vice versa.*
(9) Ability to fix the identified error and confirm the fix through testing.*

Next, for each FKSA that was to be included, we identified the evidence needed and the required task characteristics to ensure that tasks would elicit evidence for or against the selected FKSAs. All tasks were scenario-based and not tied to any particular programming representation. We paid particular attention to the scenarios to ensure that the reading load was manageable for students, the contexts were diverse and culturally relatable to students in Hong Kong, and they did not require student knowledge on concepts other than the target CT constructs being assessed. A suite of about 20 tasks was initially developed, from which 12 tasks were selected for the final CT practice assessment instrument, each with between 1 and 5 items (sub-tasks). 2 of the 12 tasks aligned with CTP1, 3 with CTP2, 3 with CTP3, 2 with CTP4, and 2 tasks assessed a combination of practices.

Figures 1–4 depict four representative assessment tasks, one corresponding to each CT Practice. Figure 1 depicts a "Reusing and remixing" task (RR1) that aligns with CTP1:FKSA1 (Ability to evaluate existing programs and problem solutions in light of a new problem or purpose). Students were assessed on whether they could identify what parts of an existing solution could be (a) reused directly, (b) reused with modification to adapt to a new context, and (c) what could not be reused at all. According to the task, Ying creates procedures or blocks of code to create a given picture. For Talia to generate another similar picture, students had to specify for each of the 6 objects in Talia's picture whether

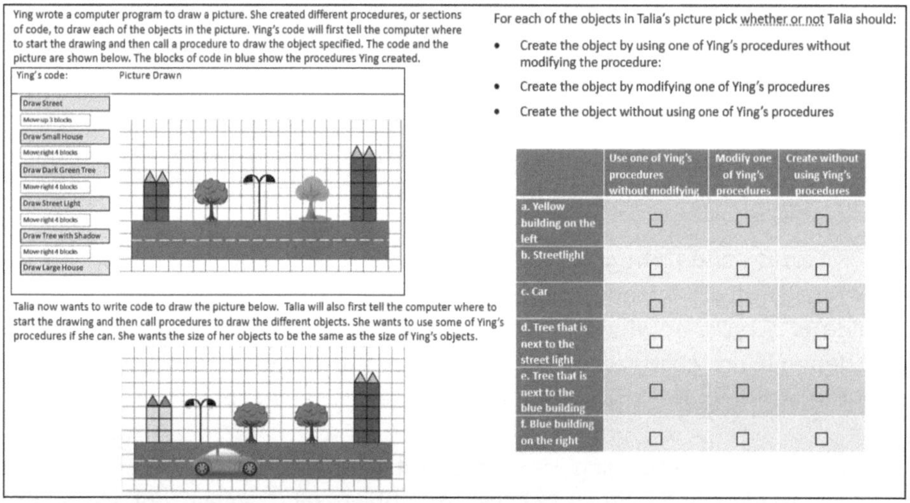
Ying wrote a computer program to draw a picture. She created different procedures, or sections of code, to draw each of the objects in the picture. Ying's code will first tell the computer where to start the drawing and then call a procedure to draw the object specified. The code and the picture are shown below. The blocks of code in blue show the procedures Ying created.

Talia now wants to write code to draw the picture below. Talia will also first tell the computer where to start the drawing and then call procedures to draw the different objects. She wants to use some of Ying's procedures if she can. She wants the size of her objects to be the same as the size of Ying's objects.

For each of the objects in Talia's picture pick whether or not Talia should:

- Create the object by using one of Ying's procedures without modifying the procedure:
- Create the object by modifying one of Ying's procedures
- Create the object without using one of Ying's procedures

	Use one of Ying's procedures without modifying	Modify one of Ying's procedures	Create without using Ying's procedures
a. Yellow building on the left	☐	☐	☐
b. Streetlight	☐	☐	☐
c. Car	☐	☐	☐
d. Tree that is next to the street light	☐	☐	☐
e. Tree that is next to the blue building	☐	☐	☐
f. Blue building on the right	☐	☐	☐

Figure 1. An example task RR1 assessing the "Reusing and remixing" CT Practice.

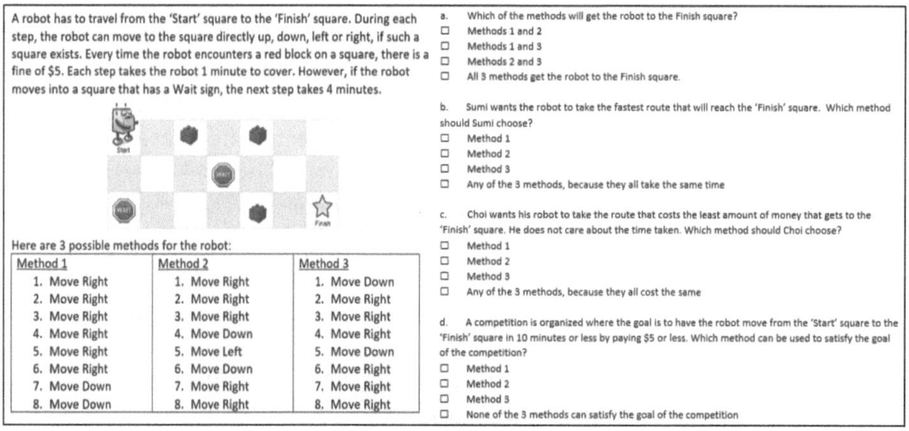
A robot has to travel from the 'Start' square to the 'Finish' square. During each step, the robot can move to the square directly up, down, left or right, if such a square exists. Every time the robot encounters a red block on a square, there is a fine of $5. Each step takes the robot 1 minute to cover. However, if the robot moves into a square that has a Wait sign, the next step takes 4 minutes.

Here are 3 possible methods for the robot:

Method 1	Method 2	Method 3
1. Move Right	1. Move Right	1. Move Down
2. Move Right	2. Move Right	2. Move Right
3. Move Right	3. Move Right	3. Move Right
4. Move Right	4. Move Down	4. Move Right
5. Move Right	5. Move Left	5. Move Down
6. Move Right	6. Move Down	6. Move Right
7. Move Down	7. Move Right	7. Move Right
8. Move Down	8. Move Right	8. Move Right

a. Which of the methods will get the robot to the Finish square?
☐ Methods 1 and 2
☐ Methods 1 and 3
☐ Methods 2 and 3
☐ All 3 methods get the robot to the Finish square.

b. Sumi wants the robot to take the fastest route that will reach the 'Finish' square. Which method should Sumi choose?
☐ Method 1
☐ Method 2
☐ Method 3
☐ Any of the 3 methods, because they all take the same time

c. Choi wants his robot to take the route that costs the least amount of money that gets to the 'Finish' square. He does not care about the time taken. Which method should Choi choose?
☐ Method 1
☐ Method 2
☐ Method 3
☐ Any of the 3 methods, because they all cost the same

d. A competition is organized where the goal is to have the robot move from the 'Start' square to the 'Finish' square in 10 minutes or less by paying $5 or less. Which method can be used to satisfy the goal of the competition?
☐ Method 1
☐ Method 2
☐ Method 3
☐ None of the 3 methods can satisfy the goal of the competition

Figure 2. An example task AT1 assessing the "Algorithmic Thinking" CT Practice.

Talia could reuse directly or remix procedures from Ying's code or whether Talia needed to create a totally new procedure.

Figure 2 describes an "Algorithmic Thinking" task (AT1) with 4 items. The task aligns with CTP2:FKSA4 (Ability to summarize the behavior of an algorithm when given a specific set of inputs) and CTP2:FKSA8 (Ability to compare the trade-offs between multiple approaches/ techniques to implementing a problem based on certain evaluation criteria). It assesses students' ability to (a) summarize the outcome or behavior of an algorithm, and (b) compare multiple approaches to solving a problem. The first item asks students which of the three given methods will take the robot from the Start to the Finish square, thus assessing students' ability to summarize the outcome of each method. The subsequent items increase the number of constraints as students are asked to compare the three methods in light of the given constraints. The second item asks students to select the method that represents the fastest route from Start to Finish, while the third item has students selecting the method

Dana and her friends are playing a "Guess the picture" game. There are 6 pictures P1-P6. Dana writes down one of the picture numbers on a paper but does not tell her friends. Now, she has to help her friends guess which picture she has selected by **giving as few hints as possible**.

Dana selected **P4**. Which of the following sets of hints should Dana give her friends that would let them pick P4 using the fewest hints possible?

a. The picture contains trees The picture contains one or more girls The picture contains balloons The picture contains a bicycle	b. The picture contains balloons The picture contains water
c. The picture contains one or more girls The picture contains balloons	d. The picture does not contain water The picture contains a bicycle The picture contains balloons

Figure 3. An example "Abstraction and Modularization" task AM1.

that incurs the least cost from Start to Finish. The fourth item combines time and cost constraints and asks students to select an option that will meet all the given criteria.

Figure 3 presents an "Abstraction and Modularization" task (AM1) that aligns with CTP3:FKSA9 (Ability to design an abstraction and evaluate the appropriateness of an abstraction for a given purpose based on the assumptions and simplifications made). Students are required to evaluate different abstractions to identify which contains fewest number of hints yet uniquely identifies a given picture and distinguishes it from other given pictures.

In Figure 4, we present a "Testing and Debugging" task (TD1) that aligns with CTP4: FKSA6 (Ability to describe a systematic method for error-detection (test cases, unit testing)) and CTP4:FKSA9 (Ability to fix the identified error and confirm the fix through testing). It assesses students' ability to identify a systematic method of error detection using meaningful test cases.

3.3 Designing a CT concepts assessment instrument

Similar to the CT practices instrument, the design of the CT concepts (programming concepts) assessment tasks also followed an ECD approach and started with determining which programming concepts were integral to the CoolThink@JC curriculum and needed to be included as part of the ECD domain analysis. Five categories of programming constructs representing the concepts of repetition, conditionals, control flow (sequences, events, parallelism), data structures (variables, lists), and procedures were selected based on the

You are testing a calculator program. The program allows you to pick two numbers and add, subtract, multiple or divide the numbers. When you ask the calculator to perform 4+2 it give you the answer of 8 instead of 6. One reason may be that the program multiplied instead of added.

What operation can you ask the calculator to perform that would let you see if the program is multiplying instead of adding?

a. $2 + 2$ *b.* 3×7 *c.* $5 + 3$ *d.* $8 - 4$

Figure 4. An example "Testing and Debugging" task TD1.

Brennan and Resnick (2012) framework. Descriptions of the concepts from this framework and related literature (Goode et al., 2012; Grover & Basu, 2017; K-12 Computer Science Framework, 2016; Maloney et al., 2008) helped obtain a deep understanding of each of these CT concept (CTC) categories. Our domain analysis also revealed important information about known student challenges related to these concepts (Basu et al., 2016; Du Boulay, 1986; Ebrahimi, 1994; Grover & Basu, 2017; Robins et al., 2003; Samurcay, 1989; Spohrer & Soloway, 1986) and helped design our assessments to ensure that our tasks overall and the distractor options in particular captured information about the presence or absence of these known challenges.

In this section, we elaborate on each of these five categories and describe how we defined corresponding FKSAs before designing tasks to elicit evidence for (or against) the FKSAs. The elaboration of the FKSAs represent the focus on specific aspects of the programming concepts as presented in the CoolThink@JC intervention; these FKSAs may need to be slightly modified to align with other curricula with different foci for these concepts.

CTC1 – Repetition:

(1) Ability to recognize or identify the output of code that includes a simple loop. This includes the ability to identify what action(s) will repeat, how long they will repeat, and what sequence they will repeat in.
(2) Ability to create a loop that sets up an action that repeats, either forever or for a set number of times.
(3) Ability to recognize or identify the output or actions of code that uses a loop to read data from a list.
(4) Ability to create a loop to read data from a list
(5) Ability to identify how often an action or set of actions occur when given an action that repeats at a set interval
(6) Ability to set up an action (or set of actions) to repeat at a set interval

CTC2 – Conditionals:

(1) Ability to identify the output of a simple conditional statement (IF-THEN or IF-THEN -ELSE)
(2) Ability to identify the output of a conditional statement that uses logical (AND, OR, NOT) and/or mathematical operators as part of the condition
(3) Ability to identify the output of nested conditional statements
(4) Ability to set up a simple conditional (IF-THEN or IF-THEN-ELSE) to represent a given narrative description
(5) Ability to set up a conditional statement that includes logical and/or mathematical operators to represent a given narrative description
(6) Ability to set up nested conditional statements to represent a given narrative description

CTC3 – Control Flow (Sequences, events, parallelism):

(1) Ability to identify the order in which actions occur in a given code segment that may include control structures such as loops and conditionals
(2) Ability to identify when variables values are updated in a given code segment that may include control structures such as loops and conditionals
(3) Ability to identify when a change in sequence of actions changes the outcome of running a code segment and when a change in sequence does not change the outcome
(4) Ability to identify two or more actions that must happen simultaneously in order to match the specifications of a given scenario
(5) Ability to identify actions that will occur at the same time in a given program
(6) Ability to create code to make two or more actions happen at the same time

CTC4 – Data Structures:

(1) Ability to identify the variable(s) that need to be used in a program and appropriate data-types for the different variables
(2) Ability to choose meaningful yet concise variable names
(3) Ability to manipulate a list (for example, add items, remove items, merge items)
(4) Ability to access items in a list and find properties of a list (for example, finding the number of items in a list, finding if a list is empty)
(5) Ability to determine the output or behavior of a program that uses a reference to a list.
(6) Ability to create multiple lists that relate to each other

CTC5 – Procedures:

(1) Ability to identify when a procedure should be called, including if it needs to be called multiple times in a program
(2) Ability to trace procedure calls and identify the set of actions that occur when a procedure call is triggered
(3) Ability to identify what input is (or should be) passed to a procedure
(4) Ability to identify the output of a procedure given the input (or calls to that procedure), including when the procedure contains a nested call to another procedure
(5) Ability to create procedures to perform specific tasks, including procedures that may have inputs and/or return outputs, and procedures that may contain nested calls to other procedures
(6) Ability to create code involving procedure calls

Since each of the three levels of the CoolThink@JC curriculum cover CT concepts of increasing complexity, the FKSAs identified for each concept also vary in their complexity. For example, CTC2:FKSA2 (Identifying the output of a conditional comprising logical or mathematical operators) and CTC2:FKSA3 (Identifying the output of nested conditionals) are evidently more complex than CTC2:FKSA1 (Identifying the output of a simple conditional). Task difficulty can also be varied by aligning tasks to multiple FKSAs instead of a single one.

We created a separate version of the CT concepts assessment instrument corresponding to each level of the curriculum. The level 1 version included a majority of tasks that are aligned to less complex FKSAs across the concepts and only a few tasks aligned to more complex FKSAs. On the other hand, the level 3 version of the assessment instrument comprised a majority of tasks aligned to complex FKSAs or a combination of FKSAs and only a few simple assessment tasks. Some tasks, both simple and complex, were common across the levels, to help evaluate students' growth over time. In addition, we varied the programming representation used in various tasks. Level 1 tasks used the Scratch programming representation or some form of pseudocode, while tasks in Levels 2 and 3 were heavily reliant on the App Inventor programming representation.

Overall, a suite of about 19 tasks was developed across the 3 levels and the four concepts of repetition, conditionals, control flow, and data structures. While "procedures" originally constituted one of the primary concept categories, revisions to the curriculum shifted the focus of the levels to the other concepts; hence, tasks assessing students' understanding of procedures were removed from the assessment instruments. 4 of the 19 tasks aligned with CTC1 (repetition), 6 with CTC2 (conditionals), 4 with CTC3 (control flow), and 5 tasks aligned with CTC4 (data structures). The CT concepts assessment instrument for each level comprised 11 tasks, each with one or more items (sub-tasks). While 5 of the 11 tasks were common across the instruments for all three levels, some tasks were common across instruments for only two levels, and some tasks were unique to the instrument for a particular level.

Figures 5–11 depict a few assessment tasks from the CT concepts assessment instrument, with varying difficulty levels, and using different programming representations. Figures 5 and 6 represent assessment tasks aligned with the CT concept of repetition. Figure 5 represents a task R1 that is aligned to CTC1:FKSA4 (Ability to create a loop to read data from a list) and uses pseudocode that is not tied to any particular programming representation. This task presents students with a desired output of animal pictures drawn a fixed distance apart and asks students to choose which set of commands would

Sia wants to write a program that will go through a list of animals and draw pictures of each animal. She wants to leave 2 blank squares between each picture of an animal. There are 7 animals in Sia's list.

The resulting picture will look like this:

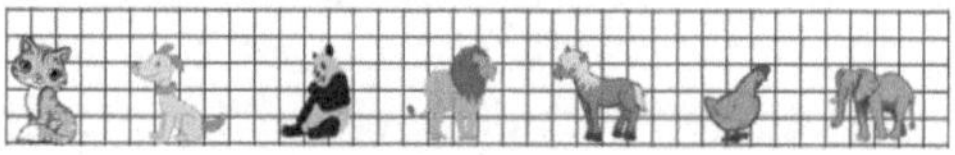

Pick the set of commands that shows how Sia can design her program to create the pictures.

☐	Let the variable *Index* = 1 Repeat the following 7 times: Get the name of the animal that is in the *Index* position on the list Draw a picture of the animal based on the name Move over 2 squares
☐	Let the variable *Index* = 1 Repeat the following 7 times Get the name of the animal that is in the *Index* position on the list Draw a picture of the animal based on the name Move over 2 squares Increase *Index* by 1
☐	Let the variable *Index* = 1 Repeat the following 2 times Get the name of the animal that is in the *Index* position on the list Draw a picture of the animal based on the name Move over 7 squares Increase *Index* by 1
☐	Let the variable *Index* = 1 Repeat the following 2 times Get the name of the animal that is in the *Index* position on the list Draw a picture of the animal based on the name Move over 7 squares Increase *Index* by 2

Figure 5. An example task R1 aligned with the CT concept of "Repetition".

The code below was created so that the *ShowItems* text box displays the items in a list.

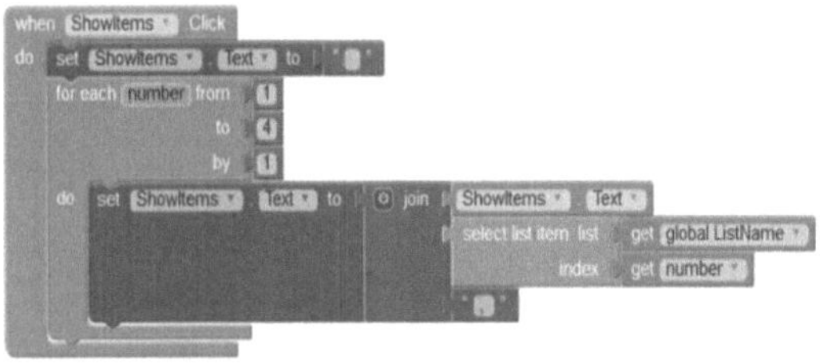

ListName is a list with the following values in it:
Chair
Table
Desk
Lamp
Rug
Bookcase

After the code is run, what text is shown in the *ShowItems* text box?

- ☐ Chair, Table, Desk, Lamp, Rug, Bookcase,
- ☐ Desk, Lamp, Rug, Bookcase,
- ☐ Lamp,
- ☐ Chair, Table, Desk, Lamp,

Figure 6. An example task R2 aligned with the CT concept of "Repetition".

correctly read data from a list to generate the desired output. Figure 6, on the other hand, presents a task R2 that aligns with CTC1:FKSA3 (Ability to recognize or identify the output or actions of code that uses a loop to read data from a list) and uses the App Inventor programming representation. Task R2 provides the contents of a list and an App Inventor code snippet to read data from the list and asks students to predict the program output.

Figure 7 depicts an assessment task C1 aligned with the CT concept of "conditionals" and CTC2:FKSA5 (Ability to set up a conditional statement that includes logical and/or mathematical operators to represent a given narrative description) that uses the App Inventor programming representation. C1 describes the age and height requirements for an amusement park ride and ask students to choose the correct conditional statement involving Boolean and mathematical operators to complete a given code snippet that helps decide eligibility for the ride.

Figures 8 and 9 present two assessment tasks aligned with the CT concept of "Control flow". Figure 8 describes a task CF1 that uses the Scratch programming representation and focuses on CTC3:FKSA1 (Ability to identify the order in which actions occur in a given code segment that may include control structures such as loops and conditionals) and CTC3:FKSA5 (Ability to identify actions that will occur at the same time in a given program). CF1 presents students with simple code snippets for two objects or sprites (dog sprite and bird sprite) involving a "broadcast" method and two events ("green flag clicked" and "dog sprite clicked"), and asks students to determine when certain actions will be made by each of the sprites. Figure 9 describes a task CF2 that also uses the Scratch programming representation but assesses the concept of "Control flow" in conjunction with the concept of "Repetition". CF2 aligns to CTC1:FKSA1 (Ability to recognize or identify the output of code that includes a simple loop.), CTC3:FKSA1 (Ability to identify the order in which actions occur in a given code segment that may include control structures such as loops and conditionals), CTC3:FKSA2 (Ability to identify when variables values are updated in a given code segment that may include control structures such as loops and conditionals), and CTC3:FKSA3 (Ability to identify when a change in sequence of actions changes the outcome of running a code segment and when a change in sequence does not change the outcome). Students are asked to compare the outputs of two given code

A water coaster ride at an amusement park has age and height requirements. You have to be at least 5 years old and at least 42 inches tall to be able to go on the ride. The following app is written to check if someone is able to take the ride.

Which of the following expressions should be attached to the *if* block?

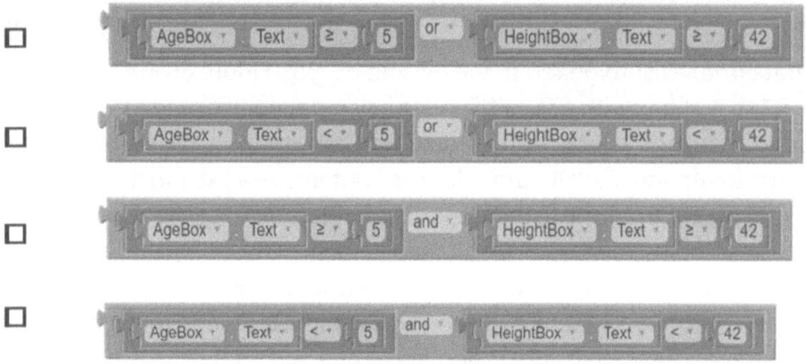

Figure 7. An example task C1 aligned with the CT concept of "Conditionals".

a. How do you make the bird move? Pick the best answer.

☐ Click the green flag

☐ Click on the bird

☐ Click on the "Next Step" button

☐ Click on the dog

b. Will the two sprites ever move at the same time?

☐ Yes, when you click the green flag

☐ Yes, when you click on the dog

☐ Yes, when you click on the bird

☐ No, they will not move at the same time.

Figure 8. An example task CF1 aligned with the CT concept of "Control Flow".

snippets involving loops and variable values that are updated within the loop structure where the difference between the two code snippets lies in the order in which the variable is updated and the variable value is recorded by the program.

Dona and Bruce write the following programs. What will the sprite say when the green flag is clicked for each of their programs?

Dona Bruce

☐ Both Dona and Bruce's programs will make the sprite say: 2, 4, 6, 8, 10

☐ Dona's program will make the sprite say: 0, 2, 4, 6, 8
while Bruce's program will make the sprite say: 0, 2, 4, 6, 8, 10

☐ Dona's program will make the sprite say: 0, 2, 4, 6, 8
while Bruce's program will make the sprite say: 2, 4, 6, 8, 10

☐ Both Dona and Bruce's programs will make the sprite say: 0, 2, 4, 6, 8

Figure 9. An example task CF2 aligned with the CT concept of "Control Flow".

Greg wants to store some information about his friends. For each type of information, decide how it should be stored.

	Text variable	Logic variable	Numeric variable	List
Name of person	o	o	o	o
Age of person	o	o	o	o
3 favorite desserts	o	o	o	o

Figure 10. An example task DS1 aligned with the CT concept of "Data structures".

Figures 10 and 11 present two assessment tasks aligned with the CT concept of "Data Structures". Figure 10 presents task DS1 aligned with CTC4:FKSA1 (Ability to identify the variable(s) that need to be used in a program and appropriate data-types for the different variables). Students are asked to choose between different data structures and data-types for storing different types of information. Figure 11 presents task DS2 that aligns to CTC4: FKSA4 (Ability to access items in a list and find properties of a list) and assesses students' understanding of the list data structure and how to access elements of a list.

3.4 Assessment piloting and validation

For both the CT practices and CT Concepts assessments, the first step in validation was an expert review. Five reviewers, experts in CS education, rated (1) the alignment of the tasks with the FKSAs, (2) the grade level appropriateness of the task and (3) the clarity and reading load of each task. For any low rating the experts provided an explanation of the rating. Reviewers were also invited to add additional comments or recommendations for

Chaoxiang creates a list, named Animal_List, of the types of animals that he has for pets. Use his list below to answer the following questions.

<u>Animal List</u>
Cat
Dog
Bunny
Snake
Hamster
Rat

a. How many values are currently included in Animal_List?
- ☐ 1
- ☐ 6
- ☐ 7
- ☐ 10
- ☐ Unknown

b. How many values would be included in the list if a new item is added to the list?
- ☐ 1
- ☐ 6
- ☐ 7
- ☐ 8
- ☐ Unknown

c. Chaoxiang uses the following code to select an item from the Animal List.

The list item that is selected is "Bunny". How should Chaoxiang change the variable named 'Position' so that the list item selected will be "Hamster"?
- ☐ Increase Position by 1
- ☐ Increase Position by 2
- ☐ Increase Position by 5
- ☐ Decrease Position by 2

Figure 11. An example task DS2 aligned with the CT concept of "Data Structures".

changes. In addition, reviewers who were Hong Kong natives provided feedback on the cultural appropriateness of the assessment task contexts for students in Hong Kong. Some other reviewers, who were part of the curriculum development team, flagged and suggested removing assessment tasks whose contexts were too close to contexts within the curriculum or difficulty levels were too high compared to what students would encounter in the curricular activities. The goal of the assessment was not to align directly with the curriculum but rather with the target FKSAs such that the assessment could be used by others targeting the same FKSAs. Tasks were revised based on the expert review to increase the clarity, readability, and alignment of each task to target FKSAs. For example, there was feedback around vocabulary terms that students might struggle with, so wording changes were made, such as using the term "program" instead of "algorithm". Another comment about cultural appropriateness of the assessment context for the target population led to removal of an item involving the algorithm for making a pizza.

Cognitive interviews were also conducted with 8 Hong Kong students to validate the assessment tasks. A native speaker (native to Hong Kong) administered the tasks and had students explain their thinking as they answered each task. Notes were taken to (1) identify any questions or challenges students had with the task, (2) determine how much time each task took, and (3) record how easy or difficult the students rated each task. Student challenges unrelated to the FKSAs (e.g., if students did not understand a phrase) were addressed by the interviewer, while challenges related to the FKSAs (e.g., when some students had difficulties comparing the algorithms, but other students did not) were simply noted. Male and female students had similar categorization of responses and comments on the tasks. A number of tasks were removed (e.g., Tower of Hanoi task for CT practices) due to their observed difficulty level or modified (task shortened and/or wording simplified) based on observations from the cognitive interviews.

For the CT practices assessment, the cognitive interviews indicated that students could engage with at most 6 tasks in one class period (45 minutes). Given the large number of tasks (12) in our instrument, a partial matrix sampling method (Childs & Jaciw, 2002) was used to distribute the tasks across three different forms. Three forms were created with three common items on each form and three unique items (see Table 1). Students were then randomly assigned (within classroom) one of the forms. The assessments were analyzed using a 1PL Item Response Theory (IRT) model to obtain scores that could be compared across students. An IRT model is a probabilistic model that generates an ability estimate for each student using the student responses and the difficulty of the questions (Hambleton et al., 1991). The use of this model allows for the creation of comparable ability estimates across students as long as there is a set of common items for the students. For overall form validation, a subset of the students (n = 1808) were given one CT practice form with all 12 tasks and given twice the amount of time as the other students. (see Table 1).

Similarly, for the CT Concepts, scores were created for each student using a 1PL IRT model. The use of one IRT model to put student performance from all three levels on the same scale was possible because the CT concepts assessment instruments were designed so that all three levels of the assessment contained a set of five common tasks (see Table 2). The number of students for each item differed depending on which form or forms the item appeared on. The total number of students who took each form was: 5698 students for level 1, 9412 students for level 2, and 1837 students for level 3. Since students in the second cohort of schools had gone through only two years of the program at the time of the analysis presented here, the number of students is lower for level 3.

We conducted validation analysis on both the CT Practices and the CT Concepts assessments for two different administrations. The results from the most recent administration in Spring 2019 are presented in Section 4. Validity evidence based on internal structure was computed using reliability and factor analysis. Additional analysis to examine item characteristics are also reported from the latest administration.

4. Analysis and results

This section discusses the results from validation activities of the assessments. Each assessment was validated separately as discussed in Section 3.4. In addition, the correlation

Table 1. Number of tasks and sample size for the CT practices assessments.

Form	Reusing and Remixing	Algorithmic Thinking	Abstraction and Modularization	Testing and Debugging	Total	Sample Size
A	1	4	1	1	7	4944
B	2	1	1	3	7	4823
C	1	2	3	1	7	5024
Validation	2	4	3	3	12	1808

Table 2. Number of tasks and total sample size for each of the CT concepts forms.

Form		Repetition	Conditionals	Control Flow	Data Structures	Total	Sample Size
Level 1	Unique to this level form	1	1	1	0	3	5698
	Shared with one other level form	0	1	1	1	3	
	On all 3 level forms	1	1	1	2	5	
Level 2	Unique to this level form	1	1	0	0	2	19,412
	Shared with one other level form	0	1	2	1	4	
	On all 3 level forms	1	1	1	2	5	
Level 3	Unique to this level form	1	2	0	2	5	1837
	Shared with one other level form	0	0	1	0	1	
	On all 3 level forms	1	1	1	2	5	

between students' ability on the CT practices assessment with ability estimate on the CT concepts assessment was calculated and found to be moderate (coefficient alpha = 0.45) indicating that while there is some relationship between scores on the two assessments, not all students who performed well on one performed well on the other. This provides an indication that these assessments were measuring different constructs and justifies the use of two separate assessment instruments for the different constructs.

4.1 CT Practices analysis and results

We established validity evidence based on internal structure by examining test reliability and factor structure for 1808 students who used the form with all 12 tasks. Each task had multiple parts and the total number of possible points was 36. The estimated Cronbach's alpha for the assessment is 0.78, meaning there is an acceptable reliability that all the items pertain to a single construct. Results of fitting a one-factor model and a four-factor model to data showed that the one-factor model fit sufficiently well (RMSEA = 0.045, with RMSEA <0.06 indicating good fit), indicating that the assessment appropriately measures one single CT practices construct. In the four-factor model, where each factor represents a CT practice, constructs correlated highly with each other (correlations ranging from .62 to .84). In addition, the distribution of scores on the assessment as a whole was as expected (see Figure 12) in that students who had received at least one year of the intervention (treatment) outperformed students of the same grade who did not have the intervention (control). In addition, for the treatment schools, students in grades Primary 5 and 6 (P5 and P6), who had more years of the intervention, outperformed the P4 students (who were exposed to only one year).

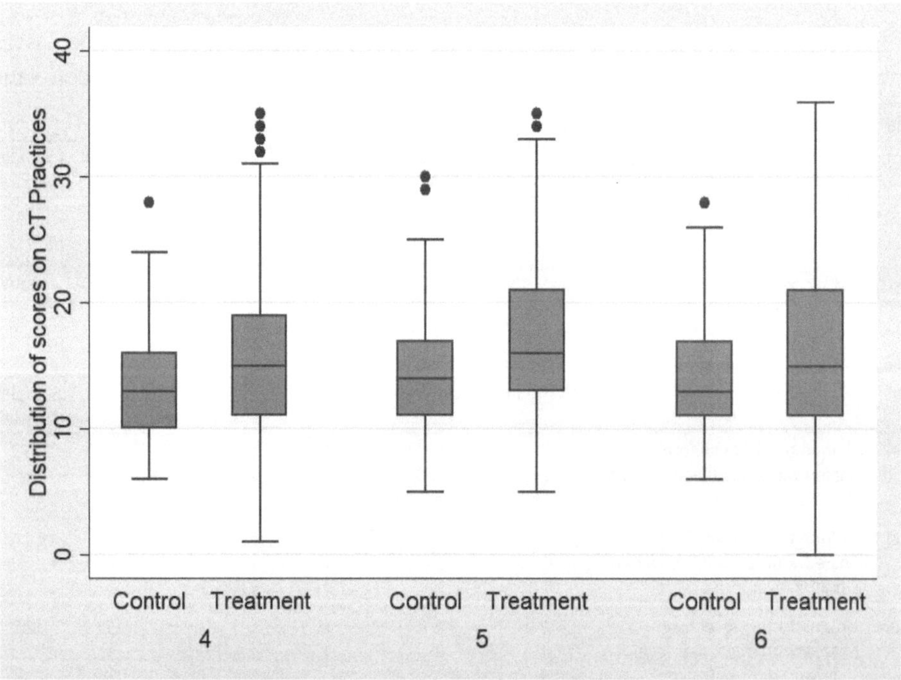

Figure 12. Distribution of student scores on the CT Concepts assessment by condition and grade level.

We also analyzed students' responses to the CT practice assessment tasks using the same 1808 students. Item analysis indicated a range of item difficulty from .16 to .82. This is considered appropriate as the assessment was designed to measure growth and therefore having a range of item difficulty was desired. In addition, a distractor analysis showed that all possible responses (distractors) were selected at least 10% of the time, indicating that the options were appropriate. We report in more detail student responses for the assessment tasks described in Section 3.2.

For CTP task RR1 described in Figure 1, we find that students could generally distinguish between procedures they could reuse from Ying's code versus procedures they would need to create anew. However, they struggled to identify which procedures they could reuse without modification versus which procedures they needed to reuse with modification to adapt to Talia's picture. For example, for the yellow building in Talia's picture, 47% students correctly identified that it could be created by reusing Ying's code with modification (modifying color), but 41% students thought it could be created by reusing without any modification. On the other hand, for objects like the streetlight and the trees in Talia's picture, about 32% of students thought that Ying's procedures needed to be modified, though the objects were identical to those in Ying's picture. Perhaps, students conflated what the object looked like with its position, though position was not a parameter for the procedure drawing the object. This might explain why most students were able to correctly identify the blue building in Talia's picture as created by "reuse without modification" – the blue building looked identical to that in Ying's picture and was also in the same location. Overall, this points to challenges with identifying and understanding procedure parameters (color can be a parameter but position is not one in this task) and recognizing what parameters or variables remain constant in a procedure versus what can change.

In CTP Task AT1 (Figure 2), several students demonstrated the ability to correctly compare the three methods (algorithms), given the various criteria. However, we notice an interesting trend – the comparison between algorithms becomes progressively more challenging for students as the number of criteria for comparison increase (Table 3). Item 1 in AT1 had only 1 criterion for comparison – moving from the Start to the Finish square. Item 2 added a second comparison criterion based on time, while Item 3 had a second criterion based on cost. Item 4 combined the Start-to-Finish, cost and time criteria, resulting in 3 criteria.

AM1 (Figure 3) was a challenging CTP task with only 22% students correctly choosing option D (Table 4). The majority of students focused on Option A that correctly identified

Table 3. Correct student responses for CTP task AT1, Items 1–4.

Comparison based on 1 criterion (Item 1)	Comparison based on 2 criteria (Item 2)	Comparison based on 2 criteria (Item 3)	Comparison based on 3 criteria (Item 4)
56.6%	50.7%	51.1%	40.3%

Table 4. Coded student responses for CTP Task AM1.

Option A (4 hints; Correctly identifies the picture)	Option B (2 hints; Points to incorrect picture)	Option C (2 hints; Points to two pictures, one of which is correct)	Option D (3 hints; Correctly identifies the picture)
34.6%	14.7%	28.6%	22.1%

the selected picture but was not the most abstract response (4 hints). The second most common distractor was Option C which included the lowest number of hints though it could not uniquely identify the selected picture. In general, students predominantly selected options A, C, or D (85.3%) that pointed to objects included in the selected picture unlike Option B that pointed to objects in other pictures. This might indicate that students find it difficult to balance the need for abstraction with the need for precisely conveying information.

In CTP task TD1 (Figure 4), students were asked to select which operation would help them test whether the program was performing multiplication instead of addition. We see that only 29% of students chose the correct option "5 + 3" while 40% of students selected "3 x 7" as their desired test case (Table 5). Students in all grades found it challenging to identify a meaningful test case that involved addition, and in particular one where addition and multiplication did not result in the same answer like in "2 + 2".

4.2 CT Concepts analysis and results

Similar analyses were conducted for each of the three CT Concepts forms (see Table 2). Results of fitting a one-factor model and a four-factor model to data showed that while the four-factor model had better model fit (an RMSEA of .051 to .057), the one-factor model also fit sufficiently well (an RMSEA of 0.061 or 0.062), indicating that it would be appropriate to generate one score from the assessments at each level. While the reliability was low for each of the three assessments (Cronbach's alpha ranged from .48 for level 3 to .66 for level 2), it is noteworthy that the results of the factor analysis were used to support use of a unidimensional model for analysis and the assessments included a mix of concepts that had been taught in the given school year and those that had not yet been taught.

The overall score distribution for assessments at each level showed a range of scores (see Figure 13), indicating that the assessments were able to differentiate between students. Each level of the form had a different possible total number of points, as the tasks differed in the number of subtasks included. Level 1 has a maximum possible number of points of 16, level 2 had 21 as the maximum, and level 3 had 19 points possible. The treatment condition (who participated in the CoolThink@JC initiative) did better than the control condition for levels 2 and 3 of the assessment. Note that while students who took level 3 were in P6, and students who took Level 1 were in P4, some students from both grades P5 and P6 took the level 2 assessment. Individual items were also reviewed to determine if the item difficulty and percent of each distractor chosen matched expectations. Item difficulties ranged from 0.24 to 0.90 and each form contained tasks that moved through this range of difficulty. Distractor analysis confirmed that students were picking distractors as expected.

Additionally, analysis of students' responses to individual tasks described in Section 3.3 helped gain insights into common student challenges, many of which have been documented in the literature on K-12 student challenges with programming. For example,

Table 5. Coded student responses for CTP Task TD1.

Option 1: 2 + 2	Option 2: 3 x 7	Option 3: 5 + 3	Option 4: 8–4
15.2%	39.8%	29.1%	15.90%

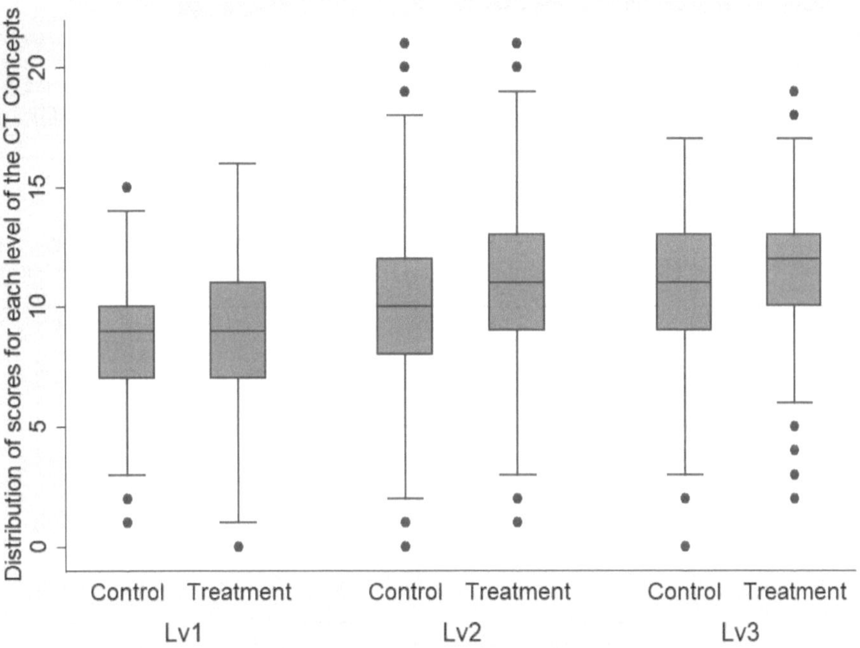

Figure 13. Distribution of student scores on each level of the CT Concepts assessment by condition.

students struggled with the "repetition" concept in combination with "variable" and "list" data structures, especially when variable values were updated within a loop and used to access different elements of a list. In the absence of variables and lists, students struggled with the order of repetition of actions when there were multiple actions inside a loop. Both these challenges bear similarities to challenges already documented in the literature (Grover & Basu, 2017; Qian et al., 2020). For task R1 (Figure 5) where students had to identify which loop structure would access a list and generate a given picture, only 27% students selected the correct option (see Table 6). The most common response (37%) was one where students correctly recognized the number of times the loop repeated but did not increase the index variable within the loop to access different list elements. This would then cause the loop to draw the same picture 7 times instead of drawing the picture of 7 different animals. For task R2 (Figure 6) where students had to follow a loop that printed out the first four items in a given list, only 24% chose the correct option. 93% students were able to recognize that the loop would print out multiple items from the list as opposed to a single list item, but they had difficulty interpreting the loop parameters and the order in which the list would be accessed. Most students (47%) stated that all the list items would be printed instead of just the first four items, while 22% students thought that the list items would be accessed bottom-up and the last 4 items would be printed.

For the CT concept of conditionals, not surprisingly, we found that students were able to correctly use simple conditional statements but had difficulty with nested conditionals and compound conditions using variables and/or logical operators. For example, only 27% students responded accurately to task C1 (Figure 7). One-third of the students chose the exact opposite of the correct response (this option would be correct if the THEN and ELSE statements were switched). Students also had challenges with the operators, with

Table 6. Student responses for CTC task R1.

Option 1: Index variable for accessing list elements is not updated	Option 2: Correct response	Option 3: Incorrect number of repetitions and incorrect spacing between pictures from the list	Option 4: Incorrect number of repetitions, incorrect spacing between pictures from the list, and incorrect increase in the value of the Index variable
36.6%	27.4%	22.6%	13.5%

Table 7. Student responses for CTC task C1.

Option 1: Uses "greater than" operator instead of "less than" operator	Option 2: Correct option	Option 3: Complement of correct option	Option 4: Uses the "AND" instead of "OR" operator
21.00%	27.14%	33.85%	18.01%

Table 8. Student responses to CTC task DS1.

	Option 1: Text variable	Option 2: Logic variable	Option 3: Numeric variable	Option 4: List
Name of the person	85.01%	4.82%	5.22%	4.92%
Age of the person	4.04%	6.03%	87.79%	2.14%
3 Favorite Desserts	9.67%	7.31%	5.16%	77.86%

some students (18%) mixing up the "AND" and the "OR" operators and other students (21%) mixing up the "greater than" and "less than" operators (see Table 7).

For Task CF1 (Figure 8) involving interpreting a Scratch program that included events and message broadcasts, most students (64%) were able to correctly identify the sequence of actions and determine that the two sprites will never move simultaneously. About half the students correctly interpreted the "broadcast" command while some students (33%) mistook the broadcast message as a button that needed to be clicked on to make the bird move. Students also found it challenging to follow the sequence of events in a code containing a loop with a variable value being updated within the loop. For example, for task CF2 (Figure 9) that aligned with the concepts of repetition and control flow, only 40% of students responded correctly (3rd option) while 66% of students recognized that there would be difference in output due to the difference in sequence of instructions within the repeat loop (2nd option + 3rd option).

As for the CT concept of "data structures" (Figures 10, 11), students were generally able to decide the appropriate type of variable or data structure to store different types of information (see Table 8). Most students (80%) were also able to determine the number of elements in a given list and how that changes if a new element is added. Half the students realized that accessing the 5th element of a list as opposed to the 3rd element requires increasing the position variable (used for accessing list elements) by 2. However, some students were unsure of how to move between elements of a list and responded that the position variable should be decreased by 2 or increased by 5.

5. Discussion and conclusion

As CT skills are increasingly emphasized in different disciplines in formal K-12 education, there is a growing need to be able to assess students on these skills in a reliable fashion.

Knowing where students are in their learning of these skills can provide teachers with valuable insights to help determine next steps in their instruction and can provide curriculum developers information to create materials that can address common student challenges (National Research Council, 2001). In this paper, we describe the design of validated CT assessment instruments for 4th-6th grade students who learn CT in the context of programming. We describe two separate instruments, one assessing students' knowledge and ability to use CT (programming) concepts and the other assessing students' generic CT (problem-solving) practices in non-programming contexts, and present results from an evaluation study where the instruments were administered to a large number of students in Hong Kong.

When developing an assessment, it is critical to clearly articulate the assessment goals and what can be said about students based on their performance. We have systematically unpacked four CT practices – Reusing and Remixing, Algorithmic Thinking, Abstraction and Modularization, and Testing and Debugging – and five CT concepts – Repetition, Conditionals, Control Flow (Sequences, Parallelism), Data Structures, and Procedures – and identified explicit learning goals or FKSAs for each of them. While we developed assessment tasks for a subset of the FKSAs, the detailed unpacking of the practices and concepts have broader implications beyond this assessment design, and can inform researchers, practitioners and curriculum developers about what it means to engage with these constructs.

Our CT concepts assessment focuses on specific programming constructs that students would have been exposed to in the CoolThink@JC curriculum and uses the Scratch and App Inventor programming representations. The CT Practices assessment, on the other hand, shifts the focus to problem-solving and is not tied to any programming representation, making the assessment useful for different contexts including CT integrated with other disciplines and CT emphasized through non-programming activities. While the choice of aspects of CT to focus on and their categorization was guided by the learning goals of the CoolThink@JC initiative, the learning goals are common enough that the assessments can be used by several other initiatives using different curricula.

Unlike several existing CT assessments that fail to meet desired psychometric standards (Cutumisu et al., 2019), results from our validation activities reveal that our assessments have psychometric properties that render them suitable to measure student learning of CT skills. The use of the principled ECD process, expert reviews and cognitive interviews provide evidence of the content validity of the assessments. Expert reviews and cognitive interviews were also used to determine if the assessments were reliant on other abilities that were not the focus of the assessment (e.g., reading load, math proficiency) and tasks were revised to minimize any such reliance. Factor analysis was used to confirm the assessment structure and support the use of an IRT model. The overall distribution of scores on the assessments and the item difficulty and distractor analysis over different cohorts provide evidence for the ability of the assessments to differentiate students with different levels of CT proficiency. It should however be noted that the assessments were validated with students in Hong Kong, and therefore may need to be validated for use in other countries and contexts. Also, with the number of items used, and the factor analysis indicating that there was one factor, it is not recommended that sub-scores for individual concepts or practices be generated based on these assessments. Further validation would be needed before using the assessments for this purpose. Additional validation examining correlations of scores on the assessments with other measures may be needed to determine the degree to which other abilities (e.g., spatial reasoning, reading ability) are related to the CT constructs being measured by the assessments. Finally, the assessments were designed to evaluate student growth over the course of the CoolThink@JC initiative and differentiate students who participated in the initiative from those who did not. Hence, the CT concepts assessment was designed to comprise tasks with a range of difficulty such that some tasks related to concepts that students had not yet encountered. Such an assessment may not be appropriate for use as an end-of-term assessment for younger students (4th or 5th grade) or as a measure of whether or not students have an understanding of basic CT concepts.

When developing any assessment, there are always trade-offs to consider, for example, between the ease/speed of scoring and the information learned from the assessment, or between the coverage of constructs in an assessment and the time needed for assessment administration. We prioritized the ease of scoring since we were working with a large number of students and balanced construct coverage and time constraints by creating multiple forms with assessment tasks distributed across forms. We chose selected response formats which may not be appropriate for assessing all learning goals; it is essential to be aware of what aspects of a construct such item types can reliably assess (Woodford & Bancroft, 2005), and what aspects are better assessed through other formats. We conclude that selected response items can be reliably used for assessing CT skills, as long as we recognize that we are only measuring specific aspects related to CT skills. In general, a combination of selected and constructed responses is ideal for probing deeper and gaining a deeper understanding of CT skills.

Additionally, we agree that process-based-measures of students' application of CT skills in dynamic programming environments can be more authentic assessment measures compared to separate static measures for CT concepts and practices using selected response formats. For example, demonstrating knowledge about a practice on a paper-pencil assessment does not necessarily equate to being able to engage in the practice while creating and testing computational artifacts in a programming environment. However, process-based measures (e.g., assessments through observations or analysis of logged student data as students are engaging in practices) may not be scalable, may be complex to interpret, and are typically not suitable as pre-post intervention measures. Just like a combination of response formats is ideal, a combination of assessment measures also helps paint a more complete picture of students' CT skills.

Finally, separate CT practices and concepts assessments help provide insights into specific student challenges in each category and point to the need for curriculum, instruction, and teacher professional development addressing these challenges. For example, our CT concepts assessment revealed that students were able to correctly create and use simple conditional statements but had difficulty with nested conditionals and conditionals using variables and logical operators. Similarly, students demonstrated proficiency on simple loops but had difficulty with loops that involved updating the value of a variable or iterating over elements of a list. For CT practices, we identified student challenges around identifying test cases for debugging and distinguishing between procedures that can be reused without modification and those that need to be modified. Encouraging a classroom culture where it is okay to make mistakes, and teaching students systematic debugging strategies, are essential but missing components in many CS classrooms. Similarly, encouraging students to reuse existing code instead of always creating their own code promotes code comprehension and impresses upon students the need for general solutions and the need to clearly comment and document their own code. Our results around student challenges with comparing algorithms when there are multiple criteria to be satisfied have implications not just for CS, but also for curricula and instruction in other disciplines such as engineering. These results help contribute to the growing literature on student learning of CT skills, and also point out the need for emphasizing CT practices in addition to CT concepts in future curricular initiatives and teacher professional development.

Acknowledgments

We gratefully acknowledge financial support for this research from the Hong Kong Jockey Club, and would like to thank current and previous collaborators who provided expert feedback on this work, helped conduct cognitive interviews in Hong Kong, and helped refine the assessment tasks, in particular, Hal Abelson, Karen Lang, Siu Cheung Kong, Mike Tissenbaum, and other colleagues from Massachusetts Institute of Technology (MIT) and The Education University of Hong Kong (EduHK).

Disclosure statement

No potential conflict of interest was reported by the authors.

ORCID

Satabdi Basu http://orcid.org/0000-0002-7055-6632
Yuning Xu http://orcid.org/0000-0002-5170-1929

References

Aho, A. V. (2012). Computation and computational thinking. *The Computer Journal*, *55*(7), 832–835. https://doi.org/10.1093/comjnl/bxs074

Alves, N. D. C., Von Wangenheim, C. G., & Hauck, J. C. (2019). Approaches to assess computational thinking competences based on code analysis in K-12 education: A systematic mapping study. *Informatics in Education*, *18*(1), 17. https://doi.org/10.15388/infedu.2019.02

Atmatzidou, S., & Demetriadis, S. (2016). Advancing students' computational thinking skills through educational robotics: A study on age and gender relevant differences. *Robotics and Autonomous Systems*, *75*(Part B), 661–670. doi:10.1016/j.robot.2015.10.008.

Barr, D., Harrison, J., & Conery, L. (2011). Computational thinking: A digital age skill for everyone. *Learning & Leading with Technology*, *38*(6), 20–23. https://files.eric.ed.gov/fulltext/EJ918910.pdf

Basu, S., Biswas, G., & Kinnebrew, J. S. (2017). Learner modeling for adaptive scaffolding in a computational thinking-based science learning environment. *User Modeling and User-Adapted Interaction*, *27*(1), 5–53. https://doi.org/10.1007/s11257-017-9187-0.

Basu, S., Biswas, G., Sengupta, P., Dickes, A., Kinnebrew, J. S., & Clark, D. (2016). Identifying middle school students' challenges in computational thinking-based science learning. *Research and Practice in Technology Enhanced Learning*, *11*(1), 13. https://doi.org/10.1186/s41039-016-0036-2

Basu, S., Rutstein, D., Xu, Y., & Shear, L. (2020). A Principled Approach to Designing a Computational Thinking Practices Assessment for Early Grades. In Proceedings of the 51st ACM Technical Symposium on Computer Science Education (pp. 912–918)

Berland, M., Martin, T., Benton, T., Petrick Smith, C., & Davis, D. (2013). Using learning analytics to understand the learning pathways of novice programmers. *Journal of the Learning Sciences*, *22*(4), 564–599. https://doi.org/10.1080/10508406.2013.836655

Bienkowski, M., Snow, E., Rutstein, D. W., & Grover, S. (2015). Assessment design patterns for computational thinking practices in secondary computer science: A first look (SRI technical report).

Blikstein, P., Worsley, M., Piech, C., Sahami, M., Cooper, S., & Koller, D. (2014). Programming pluralism: Using learning analytics to detect patterns in the learning of computer programming. *Journal of the Learning Sciences*, *23*(4), 561–599. https://doi.org/10.1080/10508406.2014.954750

Bocconi, S., Chioccariello, A., Dettori, G., Ferrari, A., Engelhardt, K., Kampylis, P., & Punie, Y. (2016). Developing computational thinking in compulsory education – Implications for policy and practice. Technical report, European Union Scientific and Technical Research Reports. EUR 28295 EN.

Brennan, K., & Resnick, M. (2012). *New frameworks for studying and assessing the development of computational thinking. Paper presented at the 2012 Annual Meeting of the American Educational Research Association*, Vancouver, Canada.

Chen, G., Shen, J., Barth-Cohen, L., Jiang, S., Huang, X., & Eltoukhy, M. (2017). Assessing elementary students' computational thinking in everyday reasoning and robotics programming. *Computers & Education*, *109*, 162–175. https://doi.org/10.1016/j.compedu.2017.03.001.

Childs, R. A., Dings, J. R., & Kingston, N. (2002). *The effect of matrix sampling on student score comparability in constructed – response and multiple- choice assessment*. Council of Chief State School Officers.

Childs, R. A., & Jaciw, A. P. (2002) Matrix sampling of items in large-scale assessments,Practical Assessment, Research, and Evaluation: Vol. 8, Article 16. https://doi.org/10.7275/gwvh-4z51

Csizmadia, A., Curzon, P., Dorling, M., Humphreys, S., Ng, T., Selby, C., & Woollard, J. (2015). Computational thinking-A guide for teachers. © Copyright 2015 Computing At School

Cutumisu, M., Adams, C., & Lu, C. (2019). A scoping review of empirical research on recent computational thinking assessments. *Journal of Science Education and Technology*, *28*(6), 651–676. https://doi.org/10.1007/s10956-019-09799-3

Dagienė, V., & Futschek, G. (2008, July). Bebras international contest on informatics and computer literacy: Criteria for good tasks. *International conference on informatics in secondary schools-evolution and perspectives* (19–30). Springer, Berlin, Heidelberg.

Dorn, B., Tew, A. E., & Guzdial, M. (2007, September). Introductory computing construct use in an end-user programming community. In *IEEE Symposium on Visual Languages and Human-Centric Computing (VL/HCC 2007)* (27–32). Coeur d'Alene, ID, USA: IEEE.

Du Boulay, B. (1986). Some difficulties of learning to program. *Journal of Educational Computing Research*, *2*(1), 57–73. https://doi.org/10.2190/3LFX-9RRF-67T8-UVK9

Ebrahimi, A. (1994). Novice programmer errors: Language constructs and plan composition. *International Journal of Human-Computer Studies*, *41*(4), 457–480. https://doi.org/10.1006/ijhc.1994.1069

Funke, J. (2012). Complex problem solving. In N. M. Seel (Ed.), *The encyclopedia of the sciences of learning* (Vol. 3, pp. 682–685). Springer; Jonassen.

Goode, J., Chapman, G., & Margolis, J. (2012). Beyond curriculum: The exploring computer science program. *ACM Inroads*, *3*(2), 47–53. https://doi.org/10.1145/2189835.2189851

Grover, S., & Basu, S. (2017, March). Measuring student learning in introductory block-based programming: Examining misconceptions of loops, variables, and boolean logic. In *Proceedings of the 2017 ACM SIGCSE technical symposium on computer science education* (267–272). Seattle Washington USA: ACM.

Grover, S., Basu, S., Bienkowski, M., Eagle, M., Diana, N., & Stamper, J. (2017). A framework for using hypothesis-driven approaches to support data-driven learning analytics in measuring computational thinking in block-based programming environments. *ACM Transactions on Computing Education (TOCE)*, *17*(3), 14.

Hambleton, R. K., Swaminathan, H., & Rogers, H. J. (1991). *Fundamentals of item response theory*. Sage.

K-12 Computer Science Framework (2016). http://www.k12cs.org

Knee, J. A., Hirsh-Pasek, K., Golinkoff, R. M., & Singer, D. (2006). *Play = learning*. Oxford University Press. https://doi.org/doi:oso/9780195304381.001.0001.

Koh, K. H., Basawapatna, A., Bennett, V., & Repenning, A. (2010, September). Towards the automatic recognition of computational thinking for adaptive visual language learning. In *2010 IEEE Symposium on Visual Languages and Human-Centric Computing* (59–66). Leganes, Spain: IEEE. doi:10.1109/VLHCC.2010.17

Kong, S. C., Abelson, H., Sheldon, J., Lao, A., Tissenbaum, M., Lai, M., Lang, K., & Lao, N. (2017, July). Curriculum activities to foster primary school students' computational practices in block-based programming environments. In *Conference Proceedings of International Conference on Computational Thinking Education* (84–89). Hong Kong: The Education University of Hong Kong.

Leonard, J., Buss, A., Gamboa, R., Mitchell, M., Fashola, O. S., Hubert, T., & Almughyirah, S. (2016). Using robotics and game design to enhance children's self-efficacy, STEM attitudes, and computational thinking skills. *Journal of Science Education and Technology*, *25*(6), 860–876. https://doi.org/10.1007/s10956-016-9628-2.

Maloney, J. H., Peppler, K., Kafai, Y., Resnick, M., & Rusk, N. (2008). Programming by choice: Urban youth learning programming with scratch. *ACM SIGCSE Bulletin*, *40*(1), 1. https://doi.org/10.1145/1352322.1352260

Martins-Pacheco, L. H., von Wangenheim, C. A. G., & Alves, N. D. C. (2019). Assessment of computational thinking in k-12 context: Educational practices, limits and possibilities-a systematic mapping study. In *Proceedings of the 11th International Conference on Computer Supported Education (CSEDU 2019)* (1, 292–303). Heraklion, Crete, Greece.

Mislevy, R. J. (2007). Validity by design. *Educational Researcher*, *36*(8), 463–469. https://doi.org/10.3102/0013189X07311660

Mislevy, R. J., & Haertel, G. D. (2006). Implications of evidence-centered design for educational testing. *Educational Measurement: Issues and Practice*, *25*(4), 6–20. https://doi.org/10.1111/j.1745-3992.2006.00075.x

Moreno-León, J., Robles, G., & Román-González, M. (2015). Dr. Scratch: Automatic analysis of scratch projects to assess and foster computational thinking. *Revista De Educacion a Distancia*, *46*, 1–23. https://www.um.es/ead/red/46/moreno_robles.pdf

Moreno-León, J., Robles, G., & Román-González, M. (2016). Code to learn: Where does it belong in the K-12 curriculum. *Journal of Information Technology Education: Research*, *15*, 283–303. https://doi.org/10.28945/3521

Moreno-León, J., Román-González, M., Harteveld, C., & Robles, G. (2017). On the automatic assessment of computational thinking skills: A comparison with human experts. In: *Proc. of the CHI Conference on Human Factors in Computing Systems*, Denver, CO, USA, 2788–2795

Mouza, C., Marzocchi, A., Pan, Y., & Pollock, L. (2016). Development, implementation, and outcomes of an equitable computer science after-school program: Findings from middle-school students. *Journal of Research on Technology in Education*, *48*(2), 84–104. https://doi.org/10.1080/15391523.2016.1146561.

National Research Council. (2001). *Knowing what students know: The science and design of educational assessment*. The National Academies Press. https://doi.org/10.17226/10019.

NGSS Lead States. (2013). *Next generation science standards: For states, by states*. National Academies Press.

Qian, Y., Hambrusch, S., Yadav, A., Gretter, S., & Li, Y. (2020). Teachers' perceptions of student misconceptions in introductory programming. *Journal of Educational Computing Research*, *58* (2), 364–397. https://doi.org/10.1177/0735633119845413

Robins, A., Rountree, J., & Rountree, N. (2003). Learning and teaching programming: A review and discussion. *Computer Science Education*, *13*(2), 137–172. https://doi.org/10.1076/csed.13.2.137.14200

Román-González, M., Pérez-González, J.-C., & Jiménez-Fernández, C. (2016). Which cognitive abilities underlie computational thinking? Criterion validity of the computational thinking test. *Computers in Human Behavior*, *72*, 678–691. https://doi.org/10.1016/j.chb.2016.08.047

Samurcay, R. (1989). The concept of variable in programming: Its meaning and use in problem-solving by novice programmers. In E. Soloway & J. C. Spohrer (Eds.), *Studying the novice programmer* (pp. 161–178). Lawrence Erlbaum Associates.

Seiter, L., & Foreman, B. (2013, August). Modeling the learning progressions of computational thinking of primary grade students. In *Proceedings of the ninth annual international ACM conference on International computing education research* (59–66). San Diego, California, USA: ACM.

Singer, D. G., Golinkoff, R. M., & Hirsh-Pasek, K. (Eds.). (2006). *Play = learning: How play motivates and enhances children's cognitive and social-emotional growth.* Oxford University Press. https://doi.org/10.1093/acprof:oso/9780195304381.001.0001

Snow, E., Rutstein, D., Basu, S., Bienkowski, M., & Everson, H. T. (2019). Leveraging Evidence-Centered Design to Develop Assessments of Computational Thinking Practices. *International Journal of Testing,19*(2), 103–127 doi:10.1080/15305058.2018.1543311

Snow, E., Rutstein, D., Bienkowski, M., & Xu, Y. (2017, August). Principled assessment of student learning in high school computer science. In *Proceedings of the 2017 ACM Conference on International Computing Education Research* (209–216). Tacoma, Washington, USA: ACM.

Spohrer, J. C., & Soloway, E. (1986). Novice mistakes: Are the folk wisdoms correct? *Communications of the ACM, 29*(7), 624–632. https://doi.org/10.1145/6138.6145

Strawhacker, A., Lee, M., & Bers, M. U. (2018). Teaching tools, teachers' rules: Exploring the impact of teaching styles on young children's programming knowledge in ScratchJr. *International Journal of Technology and Design Education, 28*(2), 347–376. https://doi.org/10.1007/s10798-017-9400-9.

Weintrop, D., & Wilensky, U. (2015, July). Using commutative assessments to compare conceptual understanding in blocks-based and text-based programs. In *ICER* (15, 101–110). Omaha, Nebraska, USA

Werner, L., Denner, J., Campe, S., & Kawamoto, D. C. (2012, February). The fairy performance assessment: Measuring computational thinking in middle school. In *Proceedings of the 43rd ACM technical symposium on Computer Science Education* (215–220). Raleigh, North Carolina, USA: ACM.

Wiebe, E., London, J., Aksit, O., Mott, B. W., Boyer, K. E., & Lester, J. C. (2019, February). Development of a lean computational thinking abilities assessment for middle grades students. In *Proceedings of the 50th ACM Technical Symposium on Computer Science Education (SIGCSE '19).* (456–461). Minneapolis, MN, USA: ACM.

Wing, J. M. (2006). Computational Thinking. *Communications of the ACM, 49*(3), 33–35. https://doi.org/10.1145/1118178.1118215

Woodford, K., & Bancroft, P., (2005) Multiple choice questions not considered harmful. *Seventh Australasian Computing Education Conference (ACE* 2005), 109–115. Darlinghurst, Australia.

Yadav, A., Burkhart, D., Moix, D., Snow, E., Bandaru, P., & Clayborn, L. (2015). *Sowing the seeds: A landscape study on assessment in secondary computer science education.* Comp. Sci. Teachers Assn.

Zur-Bargury, I., Pârv, B., & Lanzberg, D. (2013, July). A nationwide exam as a tool for improving a new curriculum. In *Proceedings of the 18th ACM conference on Innovation and technology in computer science education (ITiCSE).* (267–272). Canterbury, England, UK: ACM.

Assessing computational thinking through the lenses of functionality and computational fluency

Shari J. Metcalf, Joseph M. Reilly, Soobin Jeon, Annie Wang, Allyson Pyers, Karen Brennan and Chris Dede

ABSTRACT

Background and Context: This study looks at computational thinking (CT) assessment of programming artifacts within the context of CT integrated with science education through computational modeling.

Objective: The goal is to explore methodologies for assessment of student-constructed computational models through two lenses: functionality and conceptual fluency.

Method: This study uses data from research with EcoMOD, a 3rd grade ecosystem science curriculum in which student pairs program computational models of a beaver building a dam. Snapshots of programs for 47 student pairs collected over time are assessed.

Findings: A functionality-based rubric provided assessment of student task progress, but was less successful at correctly identifying CT gains in programs that were partially correct. A rubric for conceptual fluency identified development of fluency in CT concepts of sequencing, loops, and conditionals.

Implications: This study contributes to the literature by exploring affordances of different rubric-based CT assessments of student programs.

Introduction

The importance of computational thinking (CT) (Wing, 2006) as a learning goal has been increasingly recognized by researchers and educators (Grover & Pea, 2018; Israel et al., 2015; Lye & Koh, 2014). CT can be generally defined as a way of thinking that includes many of the skills and practices that are part of computer programming, such as algorithmic thinking, logic, abstraction, generalization, decomposition, and debugging, but are also applicable to many areas beyond programming. (Buitrago Flórez et al., 2017).

CT is recognized as a foundational competency for problem solving in STEM contexts (Grover & Pea, 2018; NGSS Lead States, 2013) Approaches to integrating computational thinking and STEM education include its use for computational modeling – using computational tools to develop models of scientific phenomena, and promoting learning through the creation, testing, and manipulation of these models (Aksit & Wiebe, 2020;

Sengupta et al., 2013; Wagh et al., 2017; Weintrop et al., 2016). Modeling is a fundamental practice of science (Giere, 1988; Lehrer, 2009; Nersessian, 2008), and CT aligns strongly with science learning as a means of representing scientific phenomena through the construction of computational models (Sengupta et al., 2013).

Assessment of CT is a growing field of research, with many forms of CT assessment developed and tested with students at different grade levels, and based on different CT definitions, models, and theoretical frameworks (Tang et al., 2020). A wide range of methodologies are used for CT assessment, including multiple-choice or open-ended tests, performance or task-based assessments, and portfolio assessments (Adams et al., 2019). Some assessments focus on CT concepts, while others include practices as well (Shute et al., 2017). The goal of embedding CT in STEM contexts brings challenges about the assessment of CT as it is integrated in different subject areas, and how those varied contexts inform the ways in which CT can be assessed (e.g. Hutchins et al., 2020).

This paper discusses methodologies to assess students' computational thinking through examining their programmed computational models. For this research, we have chosen to develop assessments of student performance on a specific task involving computational modelling as part of a 3rd grade science curriculum. Using a curriculum that involves students in developing computational models, we present two complementary methodologies for assessing the code of these programs. Each methodology acts as a distinct lens on the data and defines the understandings we are able to measure; the strategies used to analyze the data provide different information for assessing student learning. We share our findings about the aspects of computational thinking that can be assessed using each type of analysis and discuss the implications of each rubric for student assessment. The information about student errors and hurdles that emerged from our analysis was also used to inform improvements in the curriculum and teacher professional development materials that formed the basis of our study.

The data for this study are derived from an elementary curriculum called EcoMOD (Dickes et al., 2019) in which students are engaged in programming computational models of scientific phenomena. EcoMOD is an ecosystems science curriculum that blends an immersive virtual environment with an agent-based computational modeling tool to support growth in computational thinking and ecosystems science for 3rd grade learners (age 8 to 9 years old). During the 14-day curriculum unit, student pairs spent two class periods programming models of a beaver building a dam. Logfile data collected on pairs' programming activities included snapshots of student code whenever they ran their programs. For this study, we analyzed these program snapshots to assess developing fluency in computational thinking concepts.

The rest of the paper is organized as follows. We first present related work in CT, computational modeling, block-based programming, and CT assessment. We describe the goals of the study, and our methods, including the design of the EcoMOD curriculum and programming tools, the study design, the classroom research that was conducted, and the two types of methodological analyses we performed with the study data. We present our findings for each of the two types of analysis, and conclude with a discussion of the findings and insights on the design of CT assessments.

Literature review

Computational thinking and programming

Programming and CT are deeply intertwined; most CT research involves students learning programming as a means to learn the concepts and skills related to computational thinking (Israel et al., 2015; Lye & Koh, 2014). Brennan and Resnick (2012) characterize programming as a "setting for developing capacities for computational thinking," and Shute, Sun, and Asbell-Clarke note that, while CT and programming are not the same, "being able to program is one benefit of being able to think computationally" (Shute et al., 2017, p. 5).

Brennan and Resnick (2012) developed a CT framework in the context of the visual block-based programming language Scratch. It involves three key dimensions: "computational concepts" (sequences, loops, events, parallelism, conditionals, operators, and data); "computational practices" (experimenting and iterating, testing and debugging, reusing and remixing, abstracting and modularizing); and "computational perspectives" (expressing, connecting, and questioning). Rich et al. (2017) derive learning trajectories for K-8 computational thinking from the research literature, describing trajectories for three of those computational concepts: sequence, repetition (or looping), and conditionals. The K-8 learning goals for these three concepts can be summarized as (Brennan & Resnick, 2012; Rich et al., 2017):

- Sequence: A task is expressed as a sequence of actions. The order in which the actions are carried out can affect the outcome.
- Looping: Many tasks include repeated actions. Loops are a mechanism for repeating an action or sequence of actions multiple times.
- Conditionals: Different conditional states may cause different actions to be carried out. More advanced learning includes the use of multiple or overlapping conditionals, and the use of conditionals with loops, e.g. to determine when to stop a repetition.

For our research, we focus on student learning of these three core computational concepts: sequences, loops, and conditionals, during the EcoMOD programming activity.

Computational modeling for science learning

Recently, there has been substantial work that looks at integrating CT with K-12 science learning (Aksit & Wiebe, 2020; Lee et al., 2014; Ryu et al., 2015; Sengupta et al., 2013; Wagh et al., 2017; Weintrop et al., 2016), primarily in the form of *computational modeling*. Computational modeling can be considered a subset of computational thinking as applied to STEM education (Sengupta et al., 2013; Weintrop et al., 2016). And likewise, it has long been recognized that a benefit of constructing scientific models is this provides a meaningful way to learn programming (Papert & Harel, 1991).

Studying a phenomenon by constructing a model provides a means for students to organize and test their knowledge of science concepts by converting concepts and relationships into computational structures that can be executed to generate model behaviors (Sengupta et al., 2013). Decomposing a phenomenon into the steps needed to construct a model can make concepts and relationships more explicit (Hutchins et al., 2020), and the real-time animations and graphs provided by computational models offer scaffolding for interpreting and understanding the modeled phenomena, as well as a means for model validation (Sengupta et al., 2013; Yoon et al., 2015).

Block-based programming and modeling

Visual programming tools and environments have made programming much more accessible to learners, and at an earlier age (Weintrop et al., 2018). Block-based programming languages such Scratch (Resnick et al., 2009) have been a significant breakthrough for learning programming. Block-based languages are important for novice learners because programming is done by dragging and dropping program blocks together, a feature which eliminates the chance for syntax errors (Weintrop & Wilensky, 2017). Block-based interfaces engage young students in intuitive and engaging ways with code, including a palette of blocks that display the available set of commands, as well as visual cues (e.g. color, shape) that indicate block categories and usage. Further, many block-based programming environments are interpreted, rather than compiled, a feature that promotes tinkering (Maloney et al., 2010). Users can cycle between editing and running code with immediate feedback, able to quickly move blocks in and out of a workspace as they run and debug projects.

Similarly, the accessibility of block-based languages has made it possible to develop computational modeling tools for STEM learning in younger grades (e.g. Dwyer et al., 2013; Wagh et al., 2017; Aksit & Wiebe, 2020).

Agent-based modeling is a type of modeling found to be particularly effective for science learning (Dickes et al., 2016; Klopfer et al., 2005; Weintrop et al., 2016). An "agent" is a computational object that is programmed with simple rules. As the agent interacts with its environment and/or other agents, in a 2D graphical visualization, emergent, observable effects appear over time. Again, visual block-based authoring environments have been found to make agent-based programming more accessible. For example, ViMap (Sengupta et al., 2015) provides an abstraction layer over NetLogo, allowing students to focus on their modeling tasks without being overwhelmed by the NetLogo programming language syntax (Hutchins et al., 2020). Sengupta et al. (2013) describe design strategies to make agent-based modeling accessible to elementary learners using a visual programming interface, including using the right level of programming primitives, providing means to step through and visualize code execution as an aid to debugging, and making it easy to iterate through build-test cycles.

Computational thinking assessment

Numerous studies have examined strategies for the assessment of CT; overviews of the state of the field are presented in recent publications (Adams et al., 2019; Tang et al., 2020). The most common assessment strategies involve multiple-choice or open-ended tests. For example, the Computational Thinking Test (CTt) for middle school learners includes 28 items and assesses seven CT programming concepts synthesized from CT frameworks, including sequences, loops, and conditionals (Roman-Gonzalez,

Pérez-González, & Jiménez-Fernández, 2017). Some assessment tests try to minimize familiarity with specific computing platforms, such as an assessment by Chen et al. (2017) that includes pre-post items measuring both coding in robotics and reasoning in everyday events. Others, like Grover and Basu (2017), use assessment items in a specific programming language (Scratch) to measure understanding of programming constructs.

Another strategy for assessment uses performance or portfolio assessment. In some cases, researchers create tasks for students to complete and use a rubric to evaluate their work products (Franklin et al., 2017; Werner et al., 2012). In others, students are given an assessment that is task-based, such as the Bebras tasks (Dagiene & Stupuriene, 2016), designed for an international computer science contest, but also used for CT assessment (Román-González, Moreno-León et al., 2017). Sometimes, the program or portfolio produced by students is created during open-ended design activities, which are then analyzed for evidence of CT understanding (Brennan & Resnick, 2012).

When student artifacts are analyzed for understanding of CT, analysis can include a determination of what level of understanding students have achieved on specific concepts or practices. For example, the application Dr. Scratch (Moreno-León et al., 2015) performs automated portfolio analysis of Scratch projects. To assign a CT score, Dr. Scratch infers competence on seven CT concepts based on analysis of code blocks; for example, competence of flow control is scored as *basic* if the user uses sequences of blocks, *developing* for use of "repeat" and "forever", and *proficient* for use of "repeat until" blocks. Franklin et al. (2017) describe analysis of student programs through identifying milestones for student progress in demonstrating understanding of CT concepts, the milestones representing different levels of proficiency. Artifact analysis can also identify particular challenges or hurdles for students; for example, Blikstein et al. (2014) used learning analytics to identify "sink-states" where students got stuck in coding.

Some evaluations of programming artifacts also look at intermediate steps or iterations of the program as it is being composed. The advantage of looking at student programming in process is this can be a means to see students' progression while solving a problem, and to gather patterns about student learning progressions and challenges (Blikstein et al., 2014; Lane & VanLehn, 2005; Villamor, 2020; Worsley & Blikstein, 2013). For example, Troiano et al. (2019) used Dr. Scratch on snapshots of student programs captured over time, rather than just their final products, in order to analyze trajectories in student CT development.

Goals of this study

As part of our research with EcoMOD, we designed a custom block-based programming tool for agent-based programming, with which students construct agent-based models of a beaver building a dam. In this paper, we describe our assessment of the agent-based modeling programs constructed by student pairs during a study conducted using the EcoMOD curriculum. The assessment process was design-based, and we examined student artifacts through two lenses. First, we conducted an analysis based on their final programming product, using a rubric-based assessment to determine the success of their code in executing the steps for the beaver to build a dam; we considered this a measure of functionality. We found a mismatch between the scoring and our intuition as educators, particularly for assessment of partially successful models, in evaluating the pairs' level of CT understanding. We therefore developed a second analytic approach that was more faithful to student learning, using a rubric-based assessment to capture developing fluency in the use of CT constructs of sequences, loops, and conditionals. We considered this a measure of conceptual fluency complementary to our measure of functionality. Through both of these lenses, we also identified common errors and hurdles that informed the design of our student-facing and teacher professional development materials.

Research questions

Our research questions for the study are as follows:

(1) What aspects of computational thinking in EcoMOD, if any, can be assessed via a functionality rubric?
(2) What aspects of computational thinking in EcoMOD, if any, can be assessed via a conceptual fluency rubric?
(3) How can CT assessment of program artifacts inform EcoMOD curriculum design?

Methods

EcoMOD curriculum

EcoMOD is a 14-day, inquiry-based curriculum for 3rd grade that focuses on ecosystem interactions, science practices around modeling, and computational thinking. In EcoMOD, students explore a forest ecosystem and learn about behaviors and causal interactions related to a beaver building a dam. When the beaver builds a dam, it creates a pond, with cascading effects over time on the landscape and other species.

EcoMOD is centered on two computational tools: a 3D immersive virtual world and a 2D modeling tool. In the 3D world (Figure 1, left), students engage with a realistic forest ecosystem, to explore, observe, collect data, and travel in time. The 3D virtual world also includes a point-of-view (POV) tool that gives students the opportunity to "be" a beaver building a dam. The experience of being a beaver helps students learn the steps involved in building a dam, as well as avoiding the predatory wolf.

Figure 1. 3D immersive virtual ecosystem (left) and Beaver POV tool (right).

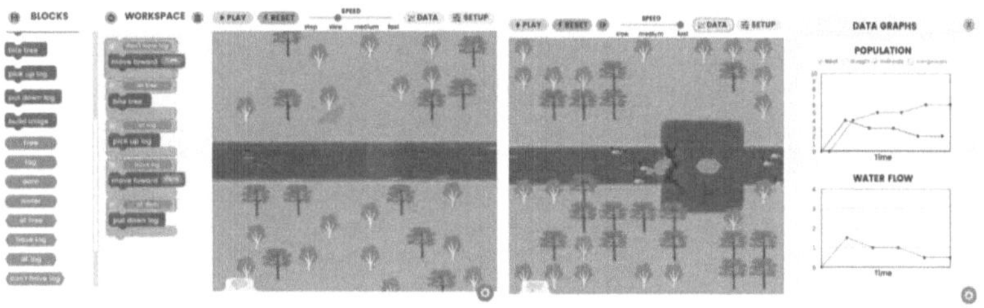

Figure 2. 2D model of a beaver building a dam (left) Data graphs in 2D tool (right).

The 2D modeling tool supports computational modeling activities, as students program an agent-based model of a beaver building a dam (Figure 2, left), using domain-specific visual programming blocks. Model outcomes are visible in a 2D sandbox that represents an abstracted version of the ecosystem (Figure 2, right). The sandbox includes other species, such as mallards and trout. These agents are not programmable but do respond to changes in the ecosystem. There are controls for different setup options (e.g. the width of the stream, the density of the trees, whether a wolf is present). Graphs allow students to view the emergent outcomes of their models.

There are four types of blocks (Figure 3): *action blocks* such as "move toward" and "bite tree", *action variables* used to specify an object (e.g. tree, log), *conditional blocks* ("if," "if/otherwise," and "repeat until"), and *conditional variables*, i.e. Boolean variables, to test state (e.g. "at tree," "have log"). The workspace has an implicit forever loop: when the program is running, the program steps through the code blocks in the workspace in a continuous loop. For example, if the "move toward tree" block is placed in the workspace and the play button is clicked, the beaver will take one step toward the closest tree, and then loop, taking another step, then another, until the beaver is at the tree and stays there.

A message window at the bottom of the screen displays feedback messages for assistance in debugging. For example, if the "bite tree" block is executed when the beaver is not at a tree, a message from the beaver says, "I am not at a tree so I can't

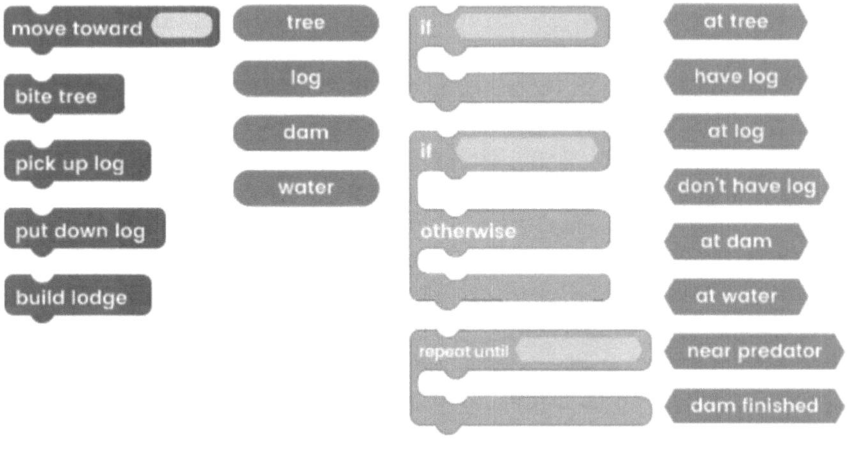

Figure 3. EcoMOD programming blocks.

Figure 4. Three successful computational models for the beaver building a dam.

bite a tree." Some errors prevent the program from running, for example, a conditional block without a variable filled in, while others (like the "bite tree" example) just skip the block's action.

Within this constrained task and limited set of blocks, there remains a range of possible working solutions. Three examples of working models developed by student pairs in this study are provided in Figure 4. In the three examples, the action blocks follow roughly the same sequence, but vary in choices of conditionals and use of nested conditionals.

Beyond a very limited introduction to the 2D modeling tool, no formal instruction in programming is included in the curriculum. The teacher explains the four types of blocks, and demonstrates the interface: how the workspace works, how to drag code blocks in and out, run the program, stop, reset, and change the speed. Students use the 2D modeling tool on days 4 to 7 of the 14-day EcoMOD curriculum. During the first two programming days, they work on programming a model of beaver building a dam, following the steps they observed and enacted in the 3D world. On the beginning of the third programming day, the teacher brings the students together to build a functional "class model" based on suggestions from students and discussion about the programming. Students who do not have a working model are encouraged to copy the class model at this point, so that they can transition to using their now functional models to explore effects of the beaver dam on the rest of the ecosystem.

Participants

This study examined data from five classes of 3rd graders (ages 8–9 years old) from three schools in three different school districts in New England, led by four different teachers. Permission slips were sent home with parents to approve participation in the study. For this study, students worked in pairs on a shared computer, with occasional exceptions when there were an odd number of students in the class or if there was a student absence. In these cases, unpaired students worked individually. After filtering for permission, data from 47 pairs were analyzed for this study. Teachers encouraged students to take turns when programming. As they engaged in coding, the teacher circulated to check in and provide support, especially for student pairs who appeared to be struggling. Students spent an average total of 45 minutes coding over two class periods, before a "class model" was built.

Data sources

Snaphots of student programs

The primary dataset for this study was obtained from the backend log files gathered automatically by the EcoMOD curriculum software and stored remotely on a University server in a PostgreSQL database. These log files contain, among other events, logged snapshots of the students' program (as in Figure 4, above) that are automatically generated each time the program is modified and run. To facilitate human coding, a script created PDF documents of these saved programs, showing each program in order, with timestamps, for each of the student pairs in the study. These program snapshots were then analyzed for this study.

Prior programming experience survey

A survey of prior experience with programming was given to the students to see what associations prior experience might have with task performance. Students were asked which common apps, languages, or websites they had used, including code.org, Scratch, Minecraft (redstone, modding, or other coding), Arduino, Javascript, Gamemaker, Lego Mindstorm, Tynker, App Creation/App Inventor, and Python, or other. Students were also asked where they had previously learned programming or coding. Then students were asked "Overall, how much programming or coding would you say that you've done?" Students who responded with "none", "less than a week," or "a few weeks" were classified as beginners, while those who selected "less than a year" or "more than a year" were classified as experienced.

Teachers formed the pairs without taking into consideration students' prior programming experience. As a result, the pair combination of beginner, intermediate, and experienced students in programming varied in each class. In this sample, 14 pairs were both beginners, 19 pairs were beginner/experienced, and 13 pairs were made up of two experienced students. Quality of programs was analyzed by prior experience level.

Rubric design and validation

Program analysis began by identifying each student pair's "final" model developed during independent programming. This was usually the last snapshot before the teacher built a model with the class. Occasionally, however, because pairs hadn't yet built a working model, they may have just cleared the workspace and started over when time was up. In those cases, beginning with the final snapshot before the class model, we worked backwards to identify the latest model that was closest to a working solution. This final model was then coded according to two different rubrics, as follows.

Functionality rubric

We first developed a *functionality rubric* to score how well the final model achieved the programming task: to build a dam and lodge while avoiding a predator. The rubric was based on the steps required of the beaver to complete the activity, as follows:

(1) Moving to a tree
(2) Biting tree until it became a log
(3) Picking up a log
(4) Bringing a log to dam and putting it down
(5) Finishing the dam (repeating the above sequence until the dam is finished)
(6) Building a lodge (executing the "build lodge" block once the dam is finished)
(7) Avoiding the predator (using the "near predator" conditional appropriately)

The task is strongly ordered, so that a program that achieves a step of the rubric must have also been able to do all the previous tasks. The final model was therefore given a score of 0–7, based on the highest-numbered step the program was able to accomplish. The main goal of the task given to the students is building a lodge, for a score of 6, but pairs who were able to build a lodge were challenged to add predator-avoidance behavior to their models as well. Models that do so successfully are given a score of 7.

The first two working models in Figure 4 (above) score 6, and the third scores 7 because it also checks for a predator nearby. But when this rubric was applied to partially built models, we found that it was less accurate at evaluating how close students were to a correct solution. For example, a very simple sequence of three blocks might score 3, as in Figure 5(a) (below), because the workspace loops automatically, so eventually the beaver will reach a tree, bite it enough times, and pick up the resulting log. Other models in which the pair shows a developing understanding of loops and conditionals, as in Figure 5 (b), score 0 since the program doesn't work at all. It fails because the beaver will take one step toward the tree, but then as the program executes the "otherwise" actions, the beaver will take one step toward the dam. This forward-and-back "dance" was a challenging hurdle for many pairs.

Conceptual fluency rubric

We then developed a *conceptual fluency* rubric that was more closely tied to student development of fluency in CT concepts. According to this rubric, the score is positively correlated with increasing complexity in the use of sequences, loops, and conditionals. The scores in this rubric represent milestones, similar to those in Franklin et al. (2017), as follows:

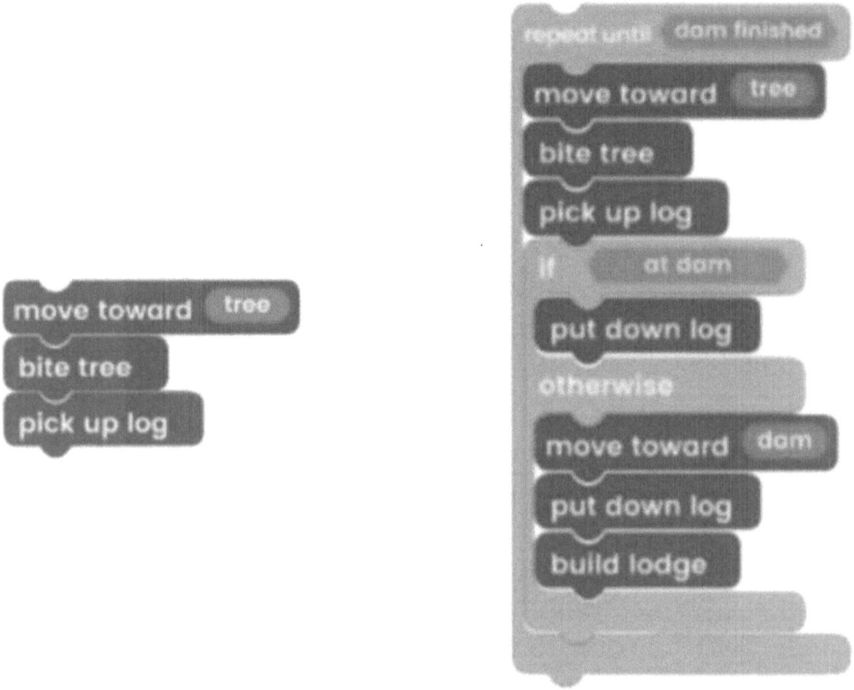

a b

Figure 5. Examples of partially built models.

(1) Sequence: putting steps in order
(2) Loops (beginning) recognizing the need to repeat instructions, evidenced one of three ways:
 (a) Using the same block multiple times to repeat an action.
 (b) Using one block at a time in the workspace, to take advantage of built-in looping.
 (c) Putting the whole sequence inside a "repeat until dam finished" block.
(3) Loops & Conditionals (basic: completing one subtask (e.g. move toward tree) using conditional blocks.
(4) Loops & Conditionals (developing): attempt to use multiple or nested conditional blocks.
(5) Loops & Conditionals (proficient): program almost successful, one error.
(6) Loops & Conditionals (advanced): Successful: fully working program.
(7) Advanced, with additional use of "predator nearby" conditional.

A score of 1, representing an understanding of sequence, involves simply putting the action blocks in the correct order, without any conditional blocks. A score of 2, beginning use of loops, includes three early strategies: (a) "step-by-step" repetition busing the same block multiple times in sequence (as in Figure 6(a)); (b) repetition "by hand," through an

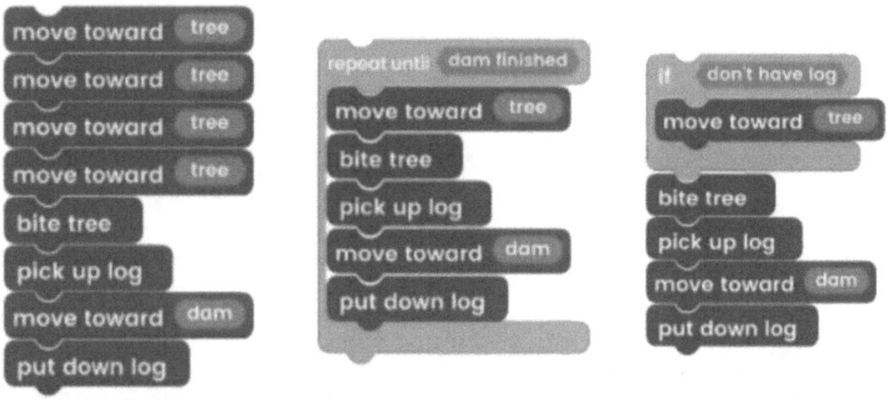

a b c

Figure 6. More examples of partially built models.

interface quirk in which code can be edited without resetting the environment: Place a single "move toward tree" block in the workspace, click play and wait until the beaver reaches a tree, then remove it and place the new block "bite tree." Then click play again, and since the beaver is already at the tree and it bites the tree, and so on, until the dam is complete; (c) placing the sequence of blocks inside a "repeat until dam finished" block (Figure 6(b)).

Scores of 3 through 6 represent stages of proficiency in the use of loops and conditionals, labeled as basic, developing, proficient, and advanced. Figure 6(c) is an example of model scoring 3, basic proficiency. Figure 5(b), above, would score 4, developing proficiency.

Rubric validation

Two raters, both members of the research team, applied the functionality rubric to all 47 pairs. The raters had prior familiarity with the software and programming task, and helped develop and refine the rubrics. Initial inter-rater reliability calculations resulted in a Cohen's Kappa of 0.631 (69% agreement). After meeting to discuss results, it was discovered that each coder had different interpretations of how to select the final program pairs used prior to being given the class model. After resolving the program selection procedure, the raters re-coded the programs with the functionality rubric with 100% agreement. Subsequently, the conceptual fluency rubric was applied in a similar fashion. Inter-rater reliability calculations for this rubric resulted in a Cohen's Kappa of 0.86 (90% agreement).

Case study comparison

In order to understand the differences found between scores using functionality and conceptual fluency rubrics, we selected two pairs to examine in-depth as a case study. One of these pairs was selected randomly from those that scored a 0 on their final functionality rubric evaluation, but had previously run programs that made more progress. The second pair was randomly chosen from those that scored a 6 on the functionality rubric. In order to explore how their programs evolve over time, both rubrics were applied to *all program snapshots* of the two pairs, in contrast with the original analyses that considered only a final model from each pair. The two pairs were from two different schools involved in this study and thus had different teachers. Pair #1 was a beginner/experienced pair; pair #2 was made up of two beginners. While more examples would be beneficial for an in-depth analysis, the development of the current rubrics for use on final student programs is the focus of the work and deploying them reliably across all programs for all students was prohibitively time-consuming.

Results

Functionality rubric

Results from applying the functionality rubric are shown in Figure 7. Scores were clustered at several points as follows. The most common score was 0, or "no tree", meaning that the program does not achieve even the basic objective of the beaver moving to a tree. The next most common stopping points were at 3, picking up a log, and at 6, completing the task by building a lodge successfully. Note that pairs scored at "avoid predator" were also able to successfully build the lodge, so 15 of the 47 pairs, or 32%, were fully successful, and 64% received a score of 3 or better.

The 11 models that scored a 0 were those in which one or more errors kept even their "best" program from working at all. Due to the continuously looping nature of the programming tool, a single "move to tree" block is enough to get the beaver to move to a tree and thus a score of 1. Once at a tree, the addition of a "bite tree" and a "pick up log" block will rapidly get pairs to the next observed plateau at a score of 3. To bring the log back to the dam, however, represents a major increase in difficulty of the task, because of the forward-and-back dance of moving toward tree and toward dam. To identify which direction to move, conditional logic using multiple conditional statements must be used

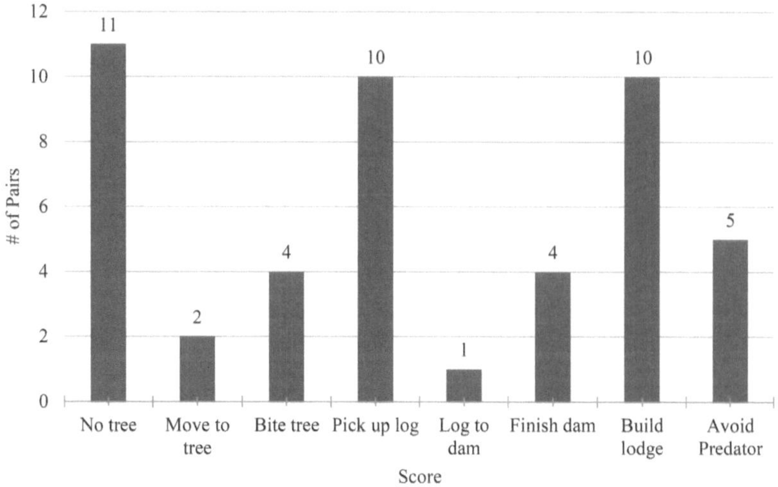

Figure 7. Count of all final functionality scores.

to successfully move on to the next scoring tier. Once that has been achieved, however, there are no new major conceptual barriers required to complete the task. Pairs that finished at "finish dam" instead of "build lodge" simply omitted a single block that actually built the lodge after the dam was complete.

No combination of prior programming experience (beginner/beginner, beginner/experienced, experienced/experienced) performed significantly better or worse on the functionality rubric based on an ANOVA model and confirmed with a Tukey's honest significant difference test. Likewise, no classes performed significantly better or worse according to this scoring, and no use of specific conditionals (if, if/otherwise, etc.) was significantly associated with higher or lower functionality rubric scores. These findings will be explicated in the discussion.

Conceptual fluency rubric

Results of coding using the conceptual fluency rubric are shown in Figure 8. 19 pairs (40%) were scored as 4, developing understanding of loops and conditionals, having recognized the need for and attempted implementation of multiple conditionals. An additional four pairs scored as proficient, with one error away from success (e.g. omitting the "build lodge" block). Scores of 6 and 7 are the same for both rubrics: ten pairs showed advanced use of loops and conditionals, with fully successful models, and five pairs additionally showed an emerging understanding of events by checking for predators while also completing other tasks.

Similar to the functionality rubric results, no combination of prior programming experience performed significantly better or worse based on an ANOVA model and confirmed with a Tukey's honest significant difference test.

Case study

For each of the two pairs selected for the in-depth case study (as described in the Methods section), we examined all program snapshots during the process of their programming

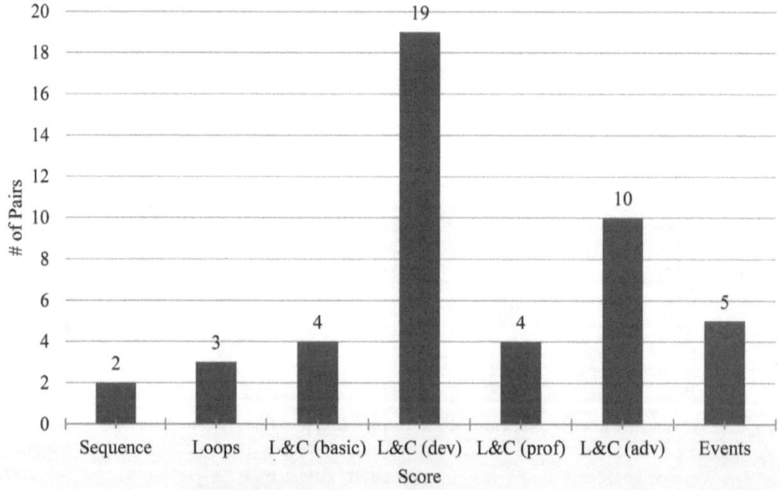

Figure 8. Count of all final conceptual fluency scores.

activities over the two class periods. Each snapshot was coded with both rubrics. Plots of how their scores change over time are provided, as well as a narrative description of how their programs changed over time. In addition to applying the rubrics, the analysis also attends to process over time by identifying the transitions between program snapshots, e.g. adding to existing programs, revising/debugging existing code, or erasing their program and starting from scratch.

Figure 9 shows a graph of the coding of the snapshots for Pair #1. This pair had 26 program snapshots, and plotlines show the change in rubric score using both the functionality (in blue) and conceptual fluency (in red) rubrics.

Pair #1 began by first running a program consisting solely of "move towards tree" then added the blocks to "bite tree" and "pick up log" in the next iteration. This second program easily accomplished the progress required for a 3 on the functionality rubric without any additional looping or conditionals. Once a log could be picked up, this pair then tried adding "move toward dam." This additional movement block in program 3 then hindered the beaver's ability to initially get to the tree, thus reducing the functionality score to 0 again. Pair #1 then tried several iterations of starting a new program and adding new blocks to in different orders during programs 3–9 to address this loss of functionality.

By program 10, this pair realized that "repeat until" blocks could solve this movement-related issue for both aspects of the task, thus raising their conceptual fluency score dramatically and regaining their prior functionality score. Programs 11–14 represent revisions to this program to fix mistakes (i.e. moving to "water" instead of "dam") and to try to implement the "put down log" behavior to the model. At program 13, the pair attempted to restructure their code to include more blocks inside "repeat until" conditions, but some errors in implementation reduced their functionality score again. Program 15 represented an attempt to start from scratch on a new day, this time utilizing "if-otherwise" blocks to achieve their desired behavior. Programs 16 through 20 represent their attempts at revising this alternate method with varying levels of success. Program 21 abandons this work in favor of a new program that attempts to include all blocks in

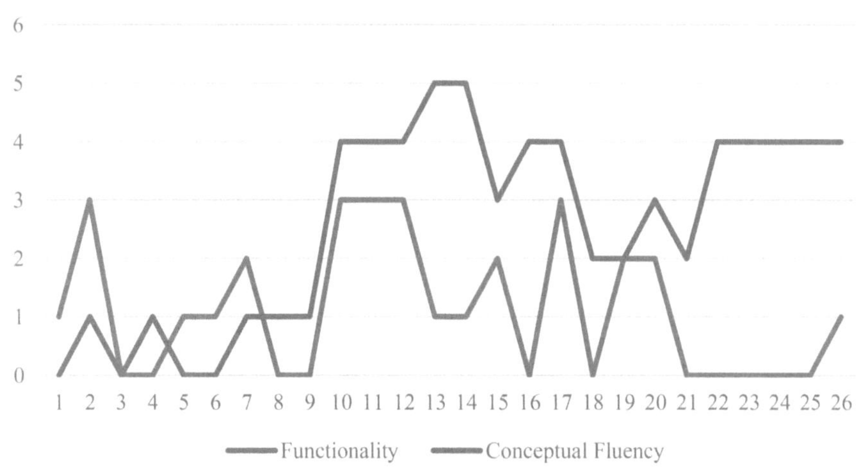

Figure 9. All functionality and conceptual fluency scores for Pair #1.

a "repeat until dam finished" conditional. In subsequent programs, multiple "repeat until" and "if" statements are nested inside this conditional during a series of revision and adding moves, but the pair never achieved the level of functionality score they achieved with a simpler program.

Pair #2, in contrast, largely achieved all functionality and conceptual fluency goals for the curriculum (Figure 10). Programs 1 and 2 established the basic sequence needed to move and bite but were missing the "tree" portion of "move toward tree." In program 3, a "repeat until at tree" conditional was added, and program 4 added behavior so that the beaver could successfully chop down a tree and pick up a log. In program 6, a second "repeat until" conditional was used to move the beaver back to the dam with a log. Program 7 showed the addition of a "put down log" command, and revisions made by program 9 allowed the beaver to successfully build the dam. This program was missing a "build lodge" command, thus limiting it from achieving a rating of 6 on the functionality rubric. After some small-scale revisions, Program 12 includes the entire program in a "repeat until dam finished" block. With the addition of a "build lodge" command after this loop in program 14, the program is thus able to successfully complete the task. An erroneously placed "move to tree" block was added in program 15 that hampered performance but was removed in the next iteration.

Discussion

Overview

With no prior instruction, fifteen of the 47 pairs were able to construct a fully functional computational model of the beaver over approximately 45 minutes spread over two class periods, and an additional four pairs were able to complete the task except the final step of building a lodge. Using the functionality rubric, we found that pairs had an average score of 3.38, with significant hurdles moving to multiple conditionals and loops that often resulted in a score of 0 for partially complete models.

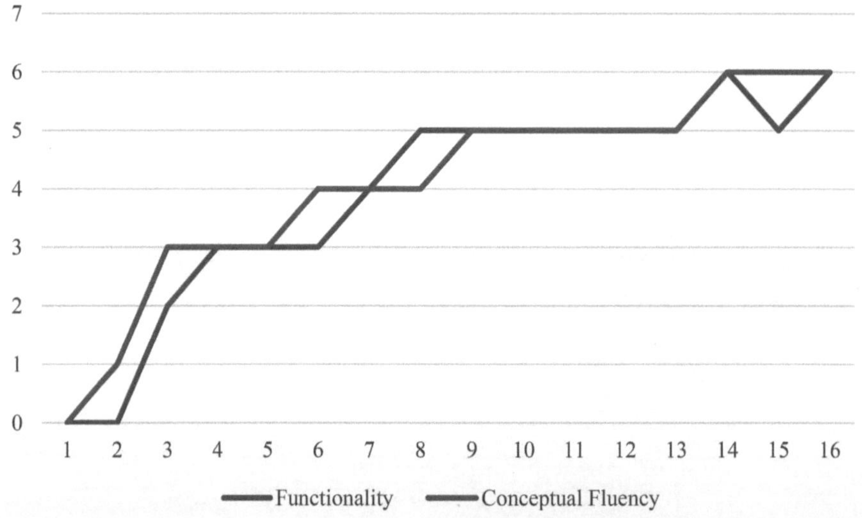

Figure 10. All functionality and conceptual fluency scores for Pair #2.

Assessment using the conceptual fluency rubric found that pairs that were not able to finish in the limited amount of time allotted still demonstrated learning of CT concepts such as sequence, loops and conditionals. All pairs showed an understanding of sequence, and almost all pairs were able to at least engage in use of multiple or nested conditionals, even when some encountered hurdles in getting them to work together correctly.

Results from both rubrics revealed no significant differences between beginner/beginner, beginner/experienced, and experienced/experienced pairs. This was a surprising finding, possibly explained by having a rather small sample, but we can also posit other possible explanations. The information on experience level was self-reported by these young students in the pre-survey, so it is possible that some misestimated the amount of time they spent programming in the past. It also seems possible that time spent in coding activities may not be the most relevant criterion, particularly considering the different types of previous experience (e.g. Minecraft redstone versus Scratch). These coding activities may have different impacts on students' abilities to generate scientific models in this curriculum, as the task might have been different enough from their prior experiences that the transfer was quite low.

The case study showed that both pairs' scores largely increased over time according to the conceptual fluency rubric as pairs grew in their use of loops and conditionals. The functionality score for pair #1, however, repeatedly rose and fell as pair #1 attempted different strategies with conditionals and wasn't able to get their code to work.

While not a focus of this research, looking at the longitudinal record of programs over time in this way also offers the opportunity to look at use of CT practices such as experimenting and iterating, testing and debugging (Brennan & Resnick, 2012). As noted in the literature, these strategies distinguish experts from novices, but are also the kinds of strategies that lead to novices becoming more effective in learning programming (Robins et al., 2003). Case study pair #1 exhibited a pattern of throwing out code that wasn't working and starting from scratch, while pair #2 iteratively added to, tested, and revised their model. Pair #2, then, could be said to be more effective in their use of CT practices, though, as noted above, pair #1 was a beginner-experienced pair and pair #2 were both beginners according to the programming pre-survey.

Assessment via functionality rubric

The first research question for this study asks what aspects of computational thinking can be assessed with a functionality rubric. This type of rubric had several advantages, namely that it provided a quick overview of student progress to see what curricular milestone pairs reached. This style of rubric is easier to understand and apply as it does not require in-depth knowledge of CT principles or learning progressions. Since the elementary teachers using this curriculum are not necessarily familiar with CT or computer science education, the scoring levels of this rubric are more easily understood by our target users.

The functionality rubric is more linked with the curricular goal of scientific modeling, and the rubric is constrained in a similar fashion to the programming tool itself. There was only one possible series of steps by which the beaver could build the dam, though there were a range of programming variations to implement those steps. A functionality rubric would be more useful to assess the scientific accuracy of the model if the code blocks were more flexible. The beaver is seen in the 3D virtual world acting a certain way and students are shown videos of real beaver behavior. These observations form the ground truth of what a beaver functionally can and cannot do, and students are instructed to mimic these behaviors in their model. If pairs could change the beaver to carry multiple logs, leap to a tree in one step, or attack the wolf, a rubric like this could assess how closely the computational model aligns with the ground truth seen in the 3D virtual world.

The results of this rubric could also inform the design of curriculum and suggest new ways of supporting learners engaged in computational modeling. By seeing when certain roadblocks happen (i.e. getting stuck trying to get the log to the dam), more supports could be added for teachers to deal with these common issues. These findings could also suggest when the class model should be introduced so that the teacher can provide support when most pairs are at a similar stage. Conversely, this rubric had the notable disadvantage of over-assigning pairs a score of 0. Even small errors in a mature program may have meant a final score of 0, thus failing to consider progress made to that point. Other very simple models resulted in higher scores since a score of 3 on the rubric could be achieved with a few simple blocks and no conditionals.

Assessment via a conceptual fluency rubric

The second research question likewise asked what aspects of computational thinking can be assessed with a conceptual fluency rubric. Unlike the functionality rubric, this rubric provides more nuanced information about student progress towards understanding CT. Students whose final model scored a 0 in the functionality rubric often scored higher here as they had made progress towards using conditionals and understood sequence at a basic level. We are able to see where they got stuck by looking at this rubric, and could identify the common hurdle of the jump to needing multiple conditional statements. Even if errors prevent programs from being fully successful, this rubric is a much better assessment of CT with a focus on learning sequences, loops, and conditionals.

This method of scoring programs can also drive curriculum development and teacher-facing support. Modeling how to properly use multiple conditionals to achieve a goal in the programming tool may help teachers see the types of errors students are making and help them understand how to intervene. Eventually, automated supports could also be included in the software depending on what error messages are being triggered.

Ramifications for the curriculum

The third research question asked how this assessment can inform curriculum design. The stages of program development exhibited by student pairs, as identified in the conceptual fluency rubric, were used to inform revisions to the design of the curriculum and professional development materials.

The computational modeling curriculum lessons were revised to address a number of issues identified through this analysis. First, a discussion of the programming blocks in the lesson plan was added to address student misconceptions. For example, some students didn't use "build lodge" because they didn't understand it, so "lodge" was introduced as a vocabulary word, as the home that beavers build in the middle of the pond once it is formed by the dam. Also added was a discussion of differences between similar blocks, e.g. What does it mean for the beaver to "have log" vs being "at log?" What is the difference between "move toward water" and "move toward dam," and when would you use each one?

Lessons during student programming were also modified to support CT processes. Recommendations were added for the teacher to model debugging strategies with the class, including using the step function to see how each block in the code is enacted, and looking at error messages to see how they can be used to debug (e.g. "I'm not at a tree so I can't bite a tree" suggests that the "bite tree" block should only be executed when the beaver is at the tree). The lesson time was extended so that students would have a chance to share their models with each other or the class – especially including opportunities to share a place where they got stuck, in order to get feedback from other students as well as the teacher.

Recommendations and tips for programming were also integrated into the teacher-facing materials, to help teachers understand and explain to students about sequences, looping, and conditionals, and to recognize common errors made by students. For each common error, the materials provide ways to identify the error, why it is an error, and ways to guide students in solving it. For example, one common error addressed in the teacher professional development materials relates to students repeating blocks instead of using a conditional (Figure 11). The "how you can help" section includes questions teachers can ask students to provide hints and encouragement for the student to figure out the next step. The step-by-step procedure will allow teachers without much programming experience to be able to support students with more confidence.

Limitations and future work

This study assesses student use of computational concepts like sequence, loops and conditionals, but it is not a direct measure of how well they understood these concepts.

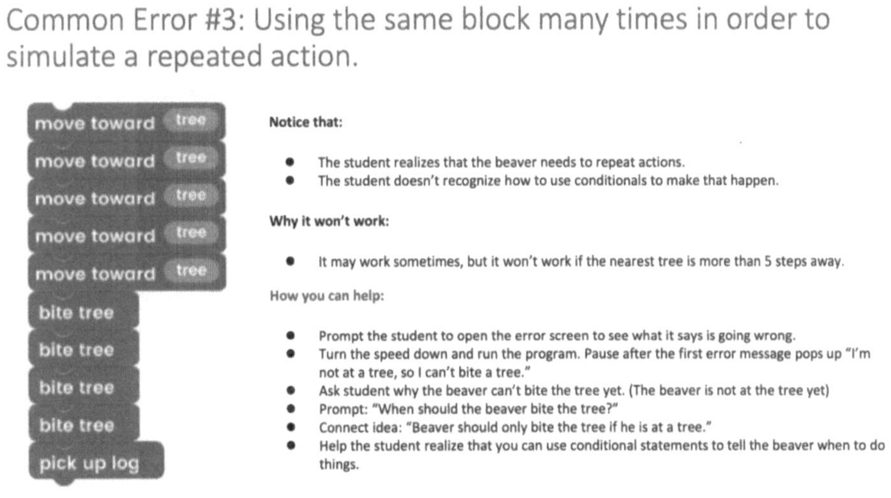

Figure 11. Example page for teachers' guide on helping learners during programming activity.

Further exploration of student learning in EcoMOD via triangulation with traditional CT assessment tests or structured interviews would help explore the extent of student CT understanding. This activity also only involves a single programming task with a specific goal. Now that the educational value of EcoMOD has been established (Dickes et al., 2019; Jeon et al., 2020), a broader curriculum with increased opportunities for modeling other types of creatures and systems may give students a chance to showcase more CT learning. Furthermore, we note that human coding of program snapshots is laborious, especially when examining partially built models that may contain errors. While necessary for eventual automation, these scoring procedures introduce a significant barrier to this type of assessment.

In addition to lowering the sample size, the use of pairs sharing one device lead to issues of not knowing how equal contributions were within pairs. The resultant pair-level scores left us unable to evaluate CT fluency by individual students within pairs. Additionally, conversations within and between pairs were common as students sat in clusters. These informal discussions are not captured and may have led to groups learning from each other. Similarly, we are unable to tell when or if teachers intervened or suggested a modification, versus pairs deciding on a new direction independently. Future work can examine classroom video to explore these interactions, to see what informal conversations looked like during the programming activity, as well as examine teacher variation in implementation.

One important area for future research would be to consider how teachers might be able to use these rubrics themselves, or to develop similar rubrics for different contexts. For this study, the research team both developed and applied the rubrics to student programs. The functionality rubric is relatively straightforward to apply; one can determine how well the program achieves the task by running it. The conceptual fluency rubric, however, requires more familiarity with coding in order to identify milestones in fluency with CT concepts, so its use by teachers would likely involve significant preparation and support.

Another interesting extension of this work could also include adapting these rubrics into automated scoring tools similar to programs like Dr. Scratch (Moreno-León et al., 2015). Further, coding of pairs' entire sequence of programs, as done for the two case study pairs in this research, might be automated and analyzed for data about student CT practices. Automated analyses might be used to identify sequences that demonstrate more mastery in practices such as iteration, testing, and debugging, and perhaps inform the design of scaffolding to guide students in these practices.

Conclusion

This study addresses CT assessment of programming artifacts constructed by student pairs to achieve a specific agent-based modeling task. We first developed a rubric based on functionality, and then, having identified limitations of this perspective, a second rubric based on identifying evidence of conceptual fluency. We found that each methodology provided a different lens on the data. The functionality rubric assessed how well the program achieved the task, but was less successful at correctly identifying CT progress in programs that were partially correct. A rubric for conceptual fluency, in contrast, was more able to recognize stages of development in use of CT concepts.

The assessment of student CT understanding with EcoMOD found that, with no direct instruction on programming, 3rd grade student pairs with minimal prior programming experience were able to make progress in using computational concepts of sequencing, loops, and conditionals. In related papers, and in our ongoing research, we examine how the design of the intervention, using a visual, block-based interface and a small set of custom, domain-specific building blocks, made the task more accessible, as did the embodiment activities in the 3D world in which students could take on the role of the beaver building the dam. (Dickes et al., 2019).

This research represents an exploration of CT assessment applied to a specific context: agent-based computational models for 3rd grade ecosystem science learning. The assessment rubrics described in the paper are based on two methodologies, functionality and computational fluency, which we suggest can be applied broadly to CT-based tasks beyond this context, to other age levels, and in different contexts. Functionality assessment focuses attention on progress towards a programming goal, while computational fluency assessment supports evaluation of milestones in use of CT concepts. It is our hope that the findings related to these two assessment strategies may inform future research and development of assessment of student programming artifacts beyond this context, and may encourage future research that considers the affordances of different dimensions of rubric-based CT assessment. We have added to the future directions section a suggestion that future work might look into how teachers might be able to use rubrics.

Acknowledgments

The authors would like to express our appreciation to Amanda Dickes for her intellectual contributions to this work. This material is based upon work supported by the National Science Foundation under grant No. DRL-1639545. Any opinions, findings and conclusions or recommendations expressed in this material are those of the authors and do not necessarily reflect the views of the National Science Foundation.

Disclosure statement

No potential conflict of interest was reported by the authors.

Funding

This work was supported by the National Science Foundation [DRL-1639545].

ORCID

Shari J. Metcalf http://orcid.org/0000-0002-7374-3249
Annie Wang http://orcid.org/0000-0001-8674-8527
Chris Dede http://orcid.org/0000-0003-0322-2461

References

Adams, C., Cutumisu, M., & Lu, C. (2019, March). Measuring K-12 computational thinking concepts, practices and perspectives: An examination of current CT assessments. In *Society for information technology & teacher education international conference* (pp. 275–285). Las Vegas, NV: Association for the Advancement of Computing in Education (AACE).

Aksit, O., & Wiebe, E. N. (2020). Exploring force and motion concepts in middle grades using computational modeling: A classroom intervention study. *Journal of Science Education and Technology, 29*(1), 65–82. https://doi.org/10.1007/s10956-019-09800-z

Blikstein, P., Worsley, M., Piech, C., Sahami, M., Cooper, S., & Koller, D. (2014). Programming pluralism: Using learning analytics to detect patterns in the learning of computer programming. *Journal of the Learning Sciences, 23*(4), 561–599. https://doi.org/10.1080/10508406.2014.954750

Brennan, K., & Resnick, M. (2012, April). New frameworks for studying and assessing the development of computational thinking. In *Proceedings of the 2012 annual meeting of the American educational research association* (Vol. 1, p. 25). Vancouver, Canada.

Buitrago Flórez, F., Casallas, R., Hernández, M., Reyes, A., Restrepo, S., & Danies, G. (2017). Changing a generation's way of thinking: Teaching computational thinking through programming. *Review of Educational Research, 87*(4), 834–860. https://doi.org/10.3102/0034654317710096

Chen, G., Shen, J., Barth-Cohen, L., Jiang, S., Huang, X., & Eltoukhy, M. (2017). Assessing elementary students' computational thinking in everyday reasoning and robotics programming. *Computers & Education, 109*(109), 162–175. https://doi.org/10.1016/j.compedu.2017.03.001

Dagiene, V., & Stupuriene, G. (2016). Bebras–A sustainable community building model for the concept based learning of informatics and computational thinking. *Informatics in Education, 15* (1), 25–44. https://doi.org/10.15388/infedu.2016.02

Dickes, A. C., Kamarainen, A., Metcalf, S. J., Gun-Yildiz, S., Brennan, K., Grotzer, T., & Dede, C. (2019). Scaffolding ecosystems science practice by blending immersive environments and computational modeling. *British Journal of Educational Technology, 50*(5), 2181–2202. https://doi.org/10.1111/bjet.12806

Dickes, A. C., Sengupta, P., Farris, A. V., & Basu, S. (2016). Development of mechanistic reasoning and multilevel explanations of ecology in third grade using agent-based models. *Science Education, 100*(4), 734–776. https://doi.org/10.1002/sce.21217

Dwyer, H., Boe, B., Hill, C., Franklin, D., & Harlow, D. (2013). Computational thinking for physics: Programming models of physics phenomenon in elementary school. In *2013 PERC (Physics Education Research Conference) Proceedings*, 133–136. https://doi.org/10.1119/perc.2013.pr.021

Franklin, D., Skifstad, G., Rolock, R., Mehrotra, I., Ding, V., Hansen, A., Weintrop, D., & Harlow, D. (2017, March). Using upper-elementary student performance to understand conceptual sequencing in a blocks-based curriculum. In *Proceedings of the 2017 ACM SIGCSE technical symposium on computer science education* (pp. 231–236). Seattle, Washington, USA: ACM.

Giere, R. N. (1988). *Explaining science: A cognitive approach*. University of Chicago Press.

Grover, S., & Basu, S. (2017, March). Measuring student learning in introductory block-based programming: Examining misconceptions of loops, variables, and boolean logic. In *Proceedings of the 2017 ACM SIGCSE technical symposium on computer science education* (pp. 267–272). Seattle, WA, USA.

Grover, S., & Pea, R. (2018). Computational thinking: A competency whose time has come. In Sentance, S., Carsten, S., & Barendsen, E. (Eds.), *Computer Science Education: Perspectives on Teaching and Learning in School*, Bloomsbury. London.

Hutchins, N. M., Biswas, G., Maróti, M., Lédeczi, Á., Grover, S., Wolf, R., Blair, K. P., Chin, D., Conlin, L., Basu, S., & McElhaney, K. (2020). C2STEM: A system for synergistic learning of physics and computational thinking. *Journal of Science Education and Technology, 29*(1), 83–100. https://doi.org/10.1007/s10956-019-09804-9

Israel, M., Pearson, J. N., Tapia, T., Wherfel, Q. M., & Reese, G. (2015). Supporting all learners in school-wide computational thinking: A cross-case qualitative analysis. *Computers & Education, 82*, 263–279. https://doi.org/10.1016/j.compedu.2014.11.022

Jeon, S., Metcalf, S., Dickes, A., & Dede, C. (2020). Elementary teacher perspectives on a blended computational modeling and ecosystem science curriculum. In D. Schmidt-Crawford (Ed.), *Proceedings of society for information technology & teacher education international conference* (pp. 46–55).

Klopfer, E., Yoon, S., & Um, T. (2005). Teaching complex dynamic systems to young students with StarLogo. *Journal of Computers in Mathematics and Science Teaching, 24*(2), 157–178. https://www.learntechlib.org/p/5537/

Lane, H. C., & VanLehn, K. (2005). Intention-based scoring: An approach to measuring success at solving the composition problem. *ACM SIGCSE Bulletin, 37*(1), 373–377. ACM. https://doi.org/10.1145/1047124.1047471

Lee, I., Martin, F., & Apone, K. (2014). Integrating computational thinking across the K--8 curriculum. *ACM Inroads, 5*(4), 64–71. https://doi.org/10.1145/2684721.2684736

Lehrer, R. (2009). Designing to develop disciplinary dispositions: Modeling natural systems. *American Psychologist, 64*(8), 759. https://doi.org/10.1037/0003-066X.64.8.759

Lye, S. Y., & Koh, J. H. L. (2014). Review on teaching and learning of computational thinking through programming: What is next for K-12? *Computers in Human Behavior, 41*, 51–61. https://doi.org/10.1016/j.chb.2014.09.012

Maloney, J., Resnick, M., Rusk, N., Silverman, B., & Eastmond, E. (2010). The scratch programming language and environment. *ACM Transactions on Computing Education (TOCE), 10*(4), 1–15. https://doi.org/10.1145/1868358.1868363

Moreno-León, J., Robles, G., & Román-González, M. (2015). Dr. Scratch: Automatic analysis of scratch projects to assess and foster computational thinking. RED. *Revista de Educación a Distancia*, (46), 1–23. https://www.um.es/ead/red/46/

Nersessian, N. (2008). Model-based reasoning in scientific practice. In *Teaching scientific inquiry* (pp. 57–79). Brill Sense. https://doi.org/10.1163/9789460911453_005

NGSS Lead States. (2013). *Next generation science standards: For states, by states*. The National Academies Press.

Papert, S., & Harel, I. (1991). Situating constructionism. *Constructionism, 36*(2), 1–11. https://web.media.mit.edu/~calla/web_comunidad/Reading-En/situating_constructionism.pdf

Resnick, M., Maloney, J., Monroy-Hernández, A., Rusk, N., Eastmond, E., Brennan, K., Millner, A., Rosenbaum, E., Silver, J., Silverman, B., & Kafai, Y. (2009). Scratch: Programming for all. *Communications of the ACM, 52*(11), 60–67. https://doi.org/10.1145/1592761.1592779

Rich, K. M., Strickland, C., Binkowski, T. A., Moran, C., & Franklin, D. (2017, August). K-8 learning trajectories derived from research literature: Sequence, repetition, conditionals. In *Proceedings of the 2017 ACM conference on international computing education research* (182–190). Tacoma, WA, USA: ACM. https://doi.org/10.1145/3105726.3106166

Robins, A., Rountree, J., & Rountree, N. (2003). Learning and teaching programming: A review and discussion. *Computer Science Education, 13*(2), 137–172. https://doi.org/10.1076/csed.13.2.137.14200

Román-González, M., Moreno-León, J., & Robles, G. (2017, July). Complementary tools for computational thinking assessment. In S. C. Kong, J. Sheldon, & K. Y. Li (Eds.). *Proceedings of international conference on Computational Thinking Education (CTE 2017)* (pp. 154–159). Hong Kong: The Education University of Hong Kong.

Román-González, M., Pérez-González, J. C., & Jiménez-Fernández, C. (2017). Which cognitive abilities underlie computational thinking? Criterion validity of the computational thinking test. *Computers in Human Behavior, 72*, 678–691. https://doi.org/10.1016/j.chb.2016.08.047

Ryu, S., Han, Y., & Paik, S. H. (2015). Understanding co-development of conceptual and epistemic understanding through modeling practices with mobile internet. *Journal of Science Education and Technology, 24*(2–3), 330–355. https://doi.org/10.1007/s10956-014-9545-1

Sengupta, P., Dickes, A., Farris, A. V., Karan, A., Martin, D., & Wright, M. (2015). Programming in K-12 science classrooms. *Communications of the ACM, 58*(11), 33–35. https://doi.org/10.1145/2822517

Sengupta, P., Kinnebrew, J. S., Basu, S., Biswas, G., & Clark, D. (2013). Integrating computational thinking with K-12 science education using agent-based computation: A theoretical framework. *Education and Information Technologies, 18*(2), 351–380. https://doi.org/10.1007/s10639-012-9240-x

Shute, V. J., Sun, C., & Asbell-Clarke, J. (2017). Demystifying computational thinking. *Educational Research Review, 22*, 142–158. https://doi.org/10.1016/j.edurev.2017.09.003

Tang, X., Yin, Y., Lin, Q., Hadad, R., & Zhai, X. (2020). Assessing computational thinking: A systematic review of empirical studies. *Computers & Education, 148*, 103798. https://doi.org/10.1016/j.compedu.2019.103798

Troiano, G. M., Snodgrass, S., Argimak, E., Robles, G., Smith, G., Cassidy, M., Tucker-Raymond, E., Puttick, G., & Harteveld, C. (2019). Is my game OK Dr. Scratch? Exploring programming and computational thinking development via metrics in student-designed serious games for STEM. In *Proceedings of the 18th ACM international conference on interaction design and children* (pp. 208–219). Boise ID, USA: ACM.

Villamor, M. M. (2020). A review on process-oriented approaches for analyzing novice solutions to programming problems. *Research and Practice in Technology Enhanced Learning, 15*(1), 1–23. https://doi.org/10.1186/s41039-020-00130-y

Wagh, A., Cook-Whitt, K., & Wilensky, U. (2017). Bridging inquiry-based science and constructionism: Exploring the alignment between students tinkering with code of computational models and goals of inquiry. *Journal of Research in Science Teaching, 54*(5), 615–641. https://doi.org/10.1002/tea.21379

Weintrop, D., Beheshti, E., Horn, M., Orton, K., Jona, K., Trouille, L., & Wilensky, U. (2016). Defining computational thinking for mathematics and science classrooms. *Journal of Science Education and Technology, 25*(1), 127–147. https://doi.org/10.1007/s10956-015-9581-5

Weintrop, D., Hansen, A. K., Harlow, D. B., & Franklin, D. (2018, August). Starting from Scratch: Outcomes of early computer science learning experiences and implications for what comes next. In *Proceedings of the 2018 ACM conference on international computing education research* (pp. 142–150). Espoo, Finland.

Weintrop, D., & Wilensky, U. (2017). Comparing block-based and text-based programming in high school computer science classrooms. *ACM Transactions on Computing Education (TOCE), 18*(1), 1–25. https://doi.org/10.1145/3089799

Werner, L., Denner, J., Campe, S., & Kawamoto, D. C. (2012, February). The fairy performance assessment: Measuring computational thinking in middle school. In *Proceedings of the 43rd ACM technical symposium on computer science education* (pp. 215–220). Raleigh, NC, USA.

Wing, J. M. (2006). Computational thinking. *Communications of the ACM, 49*(3), 33–35. https://doi.org/10.1145/1118178.1118215

Worsley, M., & Blikstein, P. (2013, July). Programming pathways: A technique for analyzing novice programmers' learning trajectories. In *International conference on artificial intelligence in education* (pp. 844–847). Berlin, Heidelberg: Springer.

Yoon, S. A., Anderson, E., Koehler-Yom, J., Klopfer, E., Sheldon, J., Wendel, D., Schoenfeld, I., Scheintaub, H., Oztok, M., & Evans, C. (2015). Design features for computer-supported complex systems learning and teaching in high school science classrooms In O. Lindwall, P. Häkkinen, T. Koschman, P. Tchounikine, & S. Ludvigsen (Eds.), *Exploring the material conditions of learning: the Computer Supported Collaborative Learning (CSCL) Conference 2015* (Vol. 1). Gothenburg, Sweden: The International Society of the Learning Sciences.

Communicating about computational thinking: understanding affordances of portfolios for assessing high school students' computational thinking and participation practices

Deborah Fields, Debora Lui, Yasmin Kafai, Gayithri Jayathirtha,
Justice Walker and Mia Shaw

ABSTRACT

Background and Context: While assessment of computational thinking concepts, practices, and perspectives is at the forefront of K-12 CS education, supporting student *communication about computation* has received relatively little attention.

Objective: To examine the usability of process-based portfolios for capturing students' communication about their computational practices regarding the process of making electronic textile projects.

Method: We examined the portfolios of 248 high school students in 15 introductory CS classrooms from largely underserved communities, using a formal rubric (top-down) to code computational communication and an open-coding scheme (bottom-up) to identify computational practices described.

Findings: Students demonstrated stronger abilities to communicate about computation using text than visuals. They also reported under-assessed CT practices like debugging, iterating, and collaborating. Students of experienced e-textile teachers performed substantially better than those with novice e-textile teachers.

Implications: Portfolios provide a viable addition to traditional performance or survey assessments and meet a need to promote communication skills.

Introduction

Computational thinking (CT), defined by Wing (2006) as "solving problems, designing systems, and understanding human behavior, by drawing on the concepts fundamental to computer science" (p. 33), has become a central concept for CS K-12 education, informing frameworks, to national standards, and curriculum design and research (e.g., K"-12 CS Framework," 2016). Yet despite its prominence, researchers and practitioners alike have not only struggled with how to define CT (National Research Council, 2010,

2011), but also how to assess it. For the most part, CT assessments have focused on computational concepts such as logic, algorithms, abstraction, generalization, evaluation, and automation (e.g., Grover, 2017; Grover & Pea, 2018), while a few others have attended to computational perspectives where students express themselves through computing or consider how computing is relevant to their lives. Much less attention has been paid to capturing computational practices such as iterative refinement, testing, and debugging (e.g., Brennan & Resnick, 2012; Grover, 2017; Grover & Pea, 2018; Jayathirtha & Kafai, 2019).

One potential avenue for assessing these under-evaluated computational practices is promoting students' abilities to communicate about computation. Despite its promotion at the college level as an essential skill in learning how to become a practitioner in the field (e.g., Falkner & Falkner, 2012; French, 2012), communication about computation has received relatively little attention in the K-12 space. Communication about computation involves describing and explaining computational artifacts and related processes and behaviors (College Board, 2017), and can be considered part of the larger category of computational participation (Kafai & Burke, 2014). This practice attends to Conley and Darling-Hammond (2013) broad call for assessments that include communication of ideas through the use of appropriate vocabulary in a subject area as well as presentation of artifacts to a broader audience. From this perspective, communicating about computation is not merely knowing the "vocabulary of computing" (Grover et al., 2014), but also understanding how to use these terms in context in order to fully engage and practice with others as part of the larger "discourse community" of the field (Gee, 1999). Developing this communication competence in computation has been shown to foster "deeper computational learning" and nurture students' abilities to both think and act upon "computational ideas more effectively" (p. 58) for the purpose of more productive learning. Notably, communicating about computation not only includes conversing with others about computation, but also documenting computational ideas and processes using both text and visual forms that are legible to a knowledgeable audience. Process-based portfolios – more commonly used in arts-based education (Gitomer et al., 1992) – provide a compelling method for practicing this skill. This format has gained traction in CS education as a result of the use of these types of portfolios in AP CS Principles as one aspect of their assessment (Arpaci-Dusseau et al., 2013; College Board, 2017).

In this paper, we report on the use of portfolios in 15 introductory computer science high school classrooms from largely underserved communities in two major metropolitan school districts. The main context of the study was the implementation of the 10–12 week electronic textiles curricular unit (Fields et al., 2018a), an optional end-of-year unit that takes place within the context of *Exploring Computer Science* (hereafter: ECS, Goode et al., 2012). Electronic textiles (hereafter: e-textiles) connect sewable microcontrollers with conductive thread to actuators such as LEDs as well as sound, light or touch sensors to make interactive craft projects. Mistakes and design changes occur frequently when students make personally relevant e-textile projects, with the implication that there is much to communicate about problems and changes that come up during the process of making e-textiles. For their portfolios, students reported on the function of a final project in the e-textile unit as well as challenges they encountered during the process of making it, namely any problems or design changes they handled. Teachers coached students on how to provide evidence of these problems or design changes, such as annotated images, to illustrate how they approached those challenges, scaffolding these reflections throughout the unit.

We examined the portfolios of 248 students to address the following research questions: RQ1: When tasked with creating a process-based portfolio about their e-textiles projects, to what extent are students able to communicate about their computational projects and processes? RQ2: What can we learn about students' engagement with computational practices, especially under-assessed activities such as testing, iteration, and debugging, through these personalized reports on process? RQ3: To what extent does prior teacher experience with implementing computational portfolios to document open-ended design affect the quality of students' communication about and reporting of their computational experience? To answer these questions, we used a rubric to evaluate the quality of students' communication quality and conducted an open-coding process to identify the range of computational thinking practices evidenced in students' portfolios. Finally, we compared the portfolios of students taught by experienced e-textile teachers (three classrooms) to those of students taught by novice e-textile teachers (twelve classrooms) both in terms of their communication quality and reports on CT practices. In the discussion, we address affordances of using portfolios to assess students' abilities to communicate about computation, as well as their engagement of computational practices in CS classrooms.

Background

One major challenge in assessing computational thinking is evaluating students' depth of understanding. Conley and Darling-Hammond (2013) argue that beyond core subject matter concepts, assessments must measure students' higher-order cognitive skills, skills that support transferable learning and also abilities like complex problem-solving, planning, reflection, collaboration and communication. For this reason, Grover (2017) proposes a "system of assessments", or multiple complementary measures that provide an expansive picture of students' learning, including cognition, affect, and transfer ability, whether evaluation of student-generated programs, specially constrained design tasks, or isolated debugging tasks (e.g., Fields et al., 2016; Moreno-Leon et al., 2015; Werner et al., 2015). Most commonly, computational assessments focus on student comprehension of computational concepts, using tools such as carefully designed quizzes, multiple choice tests, special coding tasks (like Parsons problems), or artifact-based interviews (e.g., Barron et al., 2002; Brennan & Resnick, 2012; Cooper et al., 2010; Grover, 2017; Parsons & Haden, 2006). Less common are assessments that focus primarily on computational practices and process, including active observation of students' problem solving, think aloud studies of students' debugging or design processes, or interviews about the kinds of reasoning students apply (e.g., Jayathirtha et al., 2020; Lewis, 2012). Yet amidst all of these, certain practices that Conley and Darling-Hammond (2013) call for in assessments, such as reflection, collaboration, and in particular, communication are missing. To this end, we focus on process-based portfolios for two reasons: first, as a means of facilitating student reflection and communication about computation, and second, as a a means of capturing student engagement with under-assessed computational practices such as testing and iteration, debugging, and isolation of problems.

Portfolios as assessments

Portfolios have been used across many disciplines and fields for assessment, though their actual formats and contents vary widely, each with different affordances for assessing student learning. One predominant type of portfolio, the "cumulative" portfolio, focuses on assembling a series of finished projects. Popular in art and design, this portfolio format involves curation of one's best projects over time, allowing a "showcase" of overall competency and skill (Býrgýn & Adnan, 2007). In contrast, portfolio formats derived from writing (Williams, 2002) tend to focus more on continuous documentation of students' learning and growth (Paulson et al., 1991). For instance, the "process" portfolio highlights the history of a single project, illustrating the trajectory from ideation to construction (McKay et al., 2015), while the "reflective" portfolio centers on an individual's growth or learning over a particular period of time, such as a semester or across a program (Paulson et al., 1991). These types of portfolios focus on building a narrative around one's own progress and growth, and are the basis of the portfolio assessment designed for this study.

Portfolios used in science, technology, engineering, and mathematics (STEM) domains tend to be more process-driven and reflection-focused, often dealing with the progression of a single project (Chang et al., 2015). Rather than only presenting a final project, these process-based portfolios often include a collection of different kinds of in-progress evidence compiled over a student's trajectory. This might include a range of non-textual forms of communication, such as initial prototype sketches for engineering projects (Eris, 2006), flowcharts and code throughout the development process (Higgs & Sabin, 2005), or photos of projects in process (Chang et al., 2015). Students provide textual explanations for this evidence, whether explanatory captions and annotations (Býrgýn & Adnan, 2007), articulation of underlying concepts (Phelps et al., 1997), or narrations of growth and learning (Paulson et al., 1991). These process-based portfolios offer insights into students' processes and thinking because of their situation within everyday practice, since it requires that students keep track of their ongoing work along the way (Býrgýn & Adnan, 2007).

As such, process-based portfolios are much better able to capture a more holistic view of student understanding and learning because they focus on process alongside product (Paulson et al., 1991). Here, learning and assessment can become seamlessly integrated within classroom practice (Gilman et al., 1995). For students, creating a portfolio can also provide agency in shaping one's learning over time, whether through continuous self-feedback and progress monitoring (Adams, 1998; De Fina, 1992), or purposeful opportunities for goal setting (Owings & Follo, 1992). Further, these types of portfolios highlight the actual aspects of creation and reasoning around production that are normally hidden when relying primarily on final products for assessment. In computer science, process-based, reflective portfolios can build on existing reflection and communication practices already used in some courses. For example, the practice of keeping a design notebook has already been noted as a way to become enculturated into fields of computer science and engineering in higher education (Eris, 2006). Similarly, reflective journal entries are a key practice in the equity-driven K-12 *Exploring Computer Science* curriculum as a way for students to not only share their work with others, but deepen their understanding of computation (Goode et al., 2012). Building on these reflective practices, a process-based portfolio can provide a focused means for communicating one's ongoing trajectory of learning and understanding through unpacking the process of creating a computational artifact.

Portfolios to support communication about computation

Within CS classes, portfolios offer the potential for supporting students' communication abilities, specifically their capacity to describe and explain computational artifacts and related processes and behaviors (College Board, 2017). While there have been some efforts to teach communication skills within CS courses at the university level (e.g., Falkner & Falkner, 2012; French, 2012), the most predominant effort to support computational communication in high school education is through the portfolio requirement for the recently launched AP Computer Science Principles (AP CSP) course. In the course, a portfolio supplements the standard multiple-choice exam, providing students an opportunity to show the development of their own algorithms and programmed projects. Here, they answer questions about a computational innovation they created, the tools and processes they used to create it, and the potential benefit and harm of how their creation consumes and produces data (Arpaci-Dusseau et al., 2013; College Board, 2017). The way that students are expected to accomplish this includes leveraging "accurate and precise language, notations, or visualizations" (College Board, 2017, p. 10) – in other words, focusing on both text *and* images. This dual focus in communication was part of the inspiration for our own requirement of textual and visual communication in our portfolio design.

Textual communication of computational thinking is not just a matter of vocabulary (i.e., knowing the right words to describe specific computational concepts), but also how well students can articulate and accurately capture their computational thinking practices through text and other available media (Grover & Pea, 2018). More concretely, this means that students' communication about computation can fall into different levels of written competence (Taffe, 1989). At the first level, this might encompass using a shared discourse and vocabulary of computing, such that they know the appropriate terms to describe specific ideas or procedures within the field. At the second level, students should have the capacity to employ this language to help explain larger computational systems or projects. Finally, the highest level of communicative competence emphasizes students' ability to clarify, support, and potentially defend their original computational ideas or projects to others. These features make communication a key illustration of computational participation (Kafai & Burke, 2014), which highlights the underlying social and cultural dimensions in understanding and designing computational systems where learners can effectively communicate and connect with the broader community about their creations.

Generating appropriate visual representations is another key aspect of becoming a competent computational communicator. Here, the AP CSP guidelines include not only "written and oral descriptions" as part of communication, but also how these descriptions are "supported by graphs, visualizations, and computational analysis" (College Board, 2017, p. 10). As part of their performance task assessments, they require students to create several visual aids, including annotated excerpts of code, and an original computational artifact that illustrates or represents their ideas in a "nontextual" way. Thus, beyond mere presentation of images, visual communication of computation encompasses creating and using visual evidence to further support a narrative and/or progress an argument. In general, the ability to not only interpret, but also create visualizations speaks to research that highlights the importance of doing so in order to promote greater scientific reasoning, create alternate opportunities to learn, and join the larger scientific discourse community (Ainsworth et al., 2011). Communicating effectively with visual evidence may be particularly relevant when thinking about physical computation, such as e-textiles and robotics, where fabricated objects are multidimensional and require multiple modes to convey all the elements of a project (McKay et al., 2015). Thus we chose to emphasize both textual and visual communication in the process-based portfolio we designed for the e-textile curricular unit.

Designing portfolios for computer science classes

The instances of portfolio uses in undergraduate engineering courses and high school AP CSP exams documented above inspired us to consider portfolios for assessing students' computational thinking within classroom settings to capture more robust insights into students' achievements, as well as situating assessment in more authentic, real-world contexts. Moreover, in the context of the e-textiles unit within the ECS curriculum, students not only program, but also design circuits, and construct physical textile artifacts – dimensions that are difficult to capture in a traditional written exam (for an exception, see Litts et al., 2017). A recent review of the implementation of e-textiles during the last decade revealed that while e-textiles activities were successful in broadening students' access and increasing interest in CS, there was little evidence of what they accomplished in deepening students' learning of computational concepts and practices (Jayathirtha & Kafai, 2019). Therefore, a process-based portfolio could allow students to describe their personal process, in their own words, and thus provide deeper insights into their computational understanding and practices (Brennan & Resnick, 2012; Shaw et al., 2019).

Over the course of several years, we designed, iterated, and implemented three different digital portfolio formats that were intentionally process-based. In our first effort to analyze the usefulness of portfolio assessment for computation, our participating teacher designed her own portfolio format (implemented in one classroom). This portfolio, created using an iBook format, included four sections, each focused on a domain particular to e-textiles: design, circuitry, crafting, and coding (Lui et al., 2019). While this portfolio format was successful at capturing a general sense of students' engagement with CT practices (e.g., that students debugged their code), often these descriptions were vague and did not provide enough detail to link these activities to specific computational concepts or ideas (e.g., what changes were actually made and why). Additionally, while students provided a plethora of visual evidence (e.g., code and pictures), these often lacked adequate explanation or contextualization within the narrative of the portfolio, thereby making it difficult to get a full picture of their computational understanding.

In the second version of portfolios (implemented in three classrooms), we encouraged both the reporting of process and self-reflection more explicitly, requiring students to create three sections for a digital portfolio presentation (Google Slides). These included: 1) *Product* – the final appearance and functionality of the project; 2) *Process* – any debugging or revisions that occurred during creation; and 3) *Reflection* – what they learned from the entire e-textile ECS unit. Students were also required to provide visual evidence to support these three categories, though they had the choice of what to include, whether blueprint drawings, screenshot of their code, or pictures of their projects at different stages of completion. As a result, in this iteration, students created their portfolios in personally relevant ways that acted as rich ideational resources to reflect on and construct their CS identities (Shaw et al., 2019). However, communication of their actual computational knowledge and processes were inconsistent (Lui et al., 2018). While some students were able to report about one or two of their debugging episodes with great detail, others provided over-generalized descriptions trying to encompass the entire scope of their process (e.g., resewing a part of their project, or double checking their code multiple times). Here, it was difficult to get a sense of how their computational knowledge informed these changes. Further, students' visual communication was still ineffective at supporting their narratives and arguments. As in the prior study, all students included multiple images and visualizations within their portfolios, but these were often missing annotations or captions to clarify their relationship to the textual descriptions or topic at hand.

Keeping in mind the findings from these two pilot studies, we redesigned the portfolio for the large-scale implementation of the e-textile unit reported in this study to better scaffold students in creating more effective descriptions and visualizations of abstract computational concepts and processes paper. By doing so, we were not only interested in improving the clarity of students' communication about computation but also working to increase reports about their computational practices in general. For the third version of portfolios, (i.e., the focus of this study) students again created digital portfolios (format was the teacher's choice) at the end of the e-textile unit to highlight their final project (implemented in 15 classrooms). The portfolio had the same three required sections as the previous iteration (*Product, Process*, and *Reflection*), but added more limitations in the latter two sections in order to clarify students' communication. Rather than allowing students to choose the scope of their own descriptions, we specifically asked them to detail only *two* challenges faced or changes made (including at least one that related to coding), and *one* thing they learned through the unit. Also, we explicitly outlined what an effective description should include, for example, both a problem *and* how it was resolved, or a before *and* after image of a revision (see Table 1). In providing these guidelines, we aimed to help students in producing more effective reports about their computational processes.

Further, we provided a grading rubric to support clear expectations for good computational communication, which included separate categories for textual and visual communication. Within the curriculum itself, we additionally added structured supports for portfolios, acknowledging the importance of teacher practice in scaffolding students' learning. This included two exemplar portfolios (created by the research team), and several student portfolios of varying quality from the prior year, with classroom discussion prompts on elements of effective visual communication (e.g., using visual markers,

Table 1. Summary of textual and visual requirements of the portfolio assignment. The points allotted to each section for the portfolio grading rubric are also included below.

Sections (100 total points)	Text-based Requirements (60 points total)	Visual-based Requirements (40 points total)
Description of Project (25 total points)	Documentation of final projects including its aesthetic design, physical form, and coded behaviors. (15 points)	Photos of the project including placement of all the electrical components (LEDs, sensors, and microcontroller). (10 points)
Description of Process (50 total points)	Reporting on two "process moments" including either: • two revisions made (changes and justifications); • two challenges faced (problems and solutions), or • one revision and one challenge. At least one of these moments must focus on coding/programming specifically. (30 points)	For each "process moment", including a before/after image, whether code screenshots, photos/drawings of the project, circuit diagrams, or engineering notebook or journal entries. (20 points)
Reflection on Learning (25 total points)	Discussion on one specific skill or competency gained throughout the entire e-textiles unit. (15 points)	Any kind of photos, drawings, or otherwise to help illustrate the skill and competency gained. Images from multiple projects allowed. (10 points)

captioning pictures, annotating code). In this way, students could analyze portfolios of varying quality in order to come to their own conclusions about what count as productive forms of visual and textual communication. As part of classroom activities, we also added several design notebook prompts (already a key part of the curriculum) to support students' documentation of specific errors fixed or changes made in the course of their designs. This would provide students with in-the-moment records of challenges faced or changes made, which could then be drawn upon during the process of creating their portfolios at the end of the unit. During the four-day teacher PD for the e-textiles unit, a half-day was devoted to portfolios and the portfolio rubric, where teachers evaluated prior student work and discussed strategies for supporting clear and specific computational communication through text and visuals.

Methods

Context and curriculum

This study is part of a larger, three-year effort to develop and research the electronic textiles unit of *Exploring Computer Science* (ECS). Three equity issues – broadening access, diversifying representation (by privileging non-dominant CS learners, practices, and artifacts), and deepening participation – formed the impetus for developing the e-textiles curriculum (Fields et al., 2018a). The 10–12 week e-textile unit centers around the design of four e-textile projects that build in complexity and creative freedom, helping students learn challenging concepts in computing, electronics, and crafting while also supporting personal expression and design (for design principles of the unit, see Fields et al., 2018a; for in depth descriptions of all projects see, 2018b; the unit in its final form is available at http://exploringcs.org/e-textiles). The final two projects of the unit are the most computationally complex. The third project consists of a collaborative classroom mural where pairs of students create individual pieces with a collectively chosen theme. Each piece incorporates four independently circuited LEDs sewn onto a piece of felt, utilizing two built-in switches on the Adafruit Circuit Playground microcontroller to trigger four lighting patterns. The fourth project incorporates a handmade human sensor, attached to either an existing object such as a sweatshirt or a newly created object such as a plush toy. The sensor consists of two aluminum foil conductive patches that when squeezed generate a range of data, which are programmed to correspond to different lighting effects based on how hard a user squeezed the project. The curriculum includes a 100-point rubric for the final project, equal to the number of available points for the portfolio, in order to support product and process equally as part of the system of assessments (Grover, 2017) of this e-textile unit.

Participants

In Spring 2017, the ECS-school district liaison in a large metropolitan area in California sent an invitation to the ECS-teacher listserv for educators to participate in the third and final year of the e-textiles pilot study. Teachers and school sites were chosen by the liaison for their diverse teaching styles, to maximize the variety of feedback on the curriculum. The liaison included himself as a pilot teacher. Lastly, when two teachers announced their need to take maternity leave mid-pilot, the replacement teachers were identified through snowball sampling. The final number of educators was 15, with teaching experience ranging from 3 and 37 years; most but not all also had taught ECS for several years. Twelve teachers were brand new to the e-textile unit (hereafter: "novice e-textile teachers") and three teachers had taught the e-textile unit the prior year (hereafter: "experienced e-textile teachers").[1] All teachers engaged in four days of e-textile professional development over a period of four months, where they became familiar with the curriculum by creating the projects from the unit and reflecting on the pedagogy used during professional development.

In Spring 2018, these 15 teachers implemented the e-textile unit in their ECS classes; classes ranged from 20–42 students. For participating schools, percentages of students from ethnically underrepresented groups (i.e., non-white) ranged between 72% and 99%; English as a second language between 2% and 41%; and those with free or reduced lunch between 47% and 97%. Over 430 students participated, but only 359 provided consent to collect their data for research purposes. Of these, 272 completed both the pre- and post-surveys of our larger study about the e-textiles curriculum (described in Kafai et al., 2019). For this study, we collected all available portfolios, ranging from 2–35 per teacher, and ended up with portfolios from 248 students available for analysis. Key reasons for the variable rate of portfolio return are discussed below.

Teachers differed in terms of how they implemented the portfolio assignment in their classrooms. This is part of the variation that teachers demonstrated in implementing the curriculum itself (Shaw et al., 2020). The main variation that influenced the portfolio return rate was the scheduling of the unit itself. While all teachers introduced portfolios as a part of the unit, some ended up not requiring submission of portfolios based on their own time constraints. Mostly, this was based on when teachers themselves decided when to start the unit (January – March 2018), how long they spent on it (10–13 weeks), and how many projects they covered in class. Five teachers (all novice e-textile teachers) faced a lack of time at the end of the school year and decided to do only three out of the four projects, focusing portfolios on the third (mural) project rather than the fourth (human sensor) project. Two other novice e-textile teachers allowed students to create portfolios collaboratively, whether in pairs (especially for the mural project, a pair-based project by design) or in groups of four (in the case of one teacher who, stretched for time, had students work on the human sensor project in pairs and the portfolio in fours). In classrooms where not all students submitted portfolios, teachers were focused on allowing students as much time as possible to finish their projects.

One important thing to note here about portfolio return rate is that the issues described above *only* occurred with novice e-textile teachers, who were teaching this unit and content for the first time. The experienced e-textile teachers, who had already taught this e-textiles unit *and* implemented the portfolio assignment (second iteration) in the year prior (for more information on this implementation, see Lui et al., 2018), were successfully able to complete the unit (i.e., all four projects as outlined), with nearly all of their students submitting portfolios. As will be described later, this difference is mostly attributable to prior experience with managing time for projects versus portfolios, as well as actively working to support portfolio creation throughout the design and construction process of e-textiles projects itself. Potential challenges of implementing portfolios in the classroom, as well as suggested future areas of research, are further addressed in our discussion.

Portfolio rubric

The portfolio rubric supports evaluation of students' performance on the digital portfolio. This rubric was organized into the same major sections as the portfolio: *Product* (25 points), *Process* (50 points), and *Reflection* (25 points). For each of these sections, students were evaluated on both: (1) *textual* description quality – inclusion of required information and relevant details, as well as clarity of language (60% of the score), and (2) *visual* representation quality – inclusion of relevant photos, drawings, or code excerpts, and if these were clear and easy to decipher through use of accompanying explanations, annotations/markers, or captions (40% of the score) (see Table 1 for point distribution). In order to support consistency of evaluation, we provided four-tier score categories for each subsection (e.g., for the 15 possible points for the textual description of the project, four categories of 0, 5, 10, or 15 points) with clearly laid out specifications for each category in terms of required content, description, and visuals. The full portfolio rubric is available in the e-textile unit and in the online supplemental materials (the first section of the portfolio rubric is in Figure 1). Below we describe the larger context of the e-textile unit within *Exploring Computer Science* and the participants and context of this particular study.

Data collection

We collected two forms of data including: the digital portfolios themselves and portfolio rubric evaluation scores. For the portfolios we collected, we removed all non-consenting students' portfolios, and a handful of (unopenable) corrupted portfolio files and incomplete portfolios (indicating that students started but did not complete the final portfolio).

PART I - Project Description (25 points maximum)

	15 PTS	10 PTS	5 PTS	0 PTS	POINTS EARNED
WRITTEN PART	The section contains complete explanations for the physical and functional descriptions of the project. This means that students specify the placement of the electronic components in their project and describe their four different light patterns.	The section is mostly complete but may miss some parts of the description, either the physical layout of the components or the complete functional description of the project.	The section either contains only one of the project descriptions (physical or functional) or both the sections are incomplete.	There is neither a mention of the physical or the functional description of the project.	
	10 PTS	7 PTS	3 PTS	0 PTS	
USE OF EVIDENCE	Provides relevant evidence that demonstrates both the physical layout and the technical function of the project that is made legible and understandable through appropriate textual annotation and/or visual markers (e.g., captions, labels, arrows, highlights).	Provides relevant evidence that demonstrates both the physical layout and the technical function of the project but with no textual annotation or visual markers (e.g., captions, labels, arrows, highlights) to assist with legibility or understanding.	Provides some evidence that does not match or support the claims made in the written description.	No evidence provided in this section.	
Comments on project description:					

Figure 1. Part 1 of the E-textile Portfolio Rubric: Project Description.

This left 221 portfolios from 15 classrooms, which included 204 individual portfolios and 17 group portfolios (for n = 248 students). As described above, teachers varied in terms of how they implemented the portfolios. While most teachers used the suggested digital presentation format (i.e., Google Slides, Microsoft Powerpoint) (81.4% of portfolios), some focused on alternate formats including multi-page websites (13.1%), digital text documents (i.e., Google Documents, Microsoft Word) (3.2%), and digital presentation with accompanying audio narration (i.e., a video) (2.3%). Additionally, some novice teachers focused on the pair mural project for the portfolios rather than the individual final human sensor project because of time constraints described above. For all portfolios, we tabulated their different digital formats provided and details about the projects described, which are all summarized in Table 2 below.

For portfolio rubric evaluations, we collected all available portfolio rubric evaluations from teachers. For consistency, we kept rubric evaluations only for portfolios for which we had a copy that would enable verification of grading and coding. Complete teacher-generated scores with accompanying section subscores were available for 7 of 15 classrooms (n = 99 portfolios representing 44.8% of all portfolios examined). As mentioned earlier, teachers received training on using the rubrics and evaluating sample student projects during PD. However, as with implementing the portfolio assignment, teachers varied in terms of how they graded these assignments. Some teachers used the rubric in ways that did not support data collection. Two teachers provided only the completed scores (out of 100) with no accompanying subscores (for the three sections, or text vs. visual communication), while one teacher only provided partial scores generated through peer grading in his class (a regular part of his pedagogical practice). In addition, five

Table 2. The formats of portfolios collected, as well as details about the projects described in the portfolios.

Medium of Portfolio	E-Textile Project	Form of Object	Genre of Object
Digital Slide Presentation: 81.4% Webpage: 13.1% Text Document: 3.2% Narrated Slides (voice-over video): 2.3%	Human Sensor Project: 79.2% Mural Project: 20.8%	Existing Artifact (e.g., a t-shirt bought at a store): 40.3% Original Artifact (e.g., handmade stuffed animal): 51.6% Unknown: 8.1%	Plush Toy: 49.3% Flat Object (e.g., banner, wall-hanging) 26.7% Clothing: 13.6% Accessory (e.g., hat): 8.1% Unknown: 2.3%

(novice) teachers provided no scores at all because of the time constraints described earlier. As such, one member of our research team (Author1) used the rubric to grade the portfolios from those classrooms (n = 122 portfolios representing 55.2% of all portfolios examined). Again, as with portfolio collection, potential teacher implementation issues regarding rubric evaluations and suggested areas for future research are further described in our discussion.

Data analysis

For this analysis, we implemented a mixed-methods approach. In order to answer RQ1 regarding the quality of students' computational communication skills in submitted portfolios, we looked at the grading rubric scores of student portfolios as evaluated by the teachers and the first author, calculating descriptive statistics of the different score categories for each portfolio. In order to answer RQ2 regarding students' reports of their engagement with computational practices, we looked at the content of digital portfolios. Here, we developed a grounded coding scheme through a two-step, open-coding process (Charmaz, 2006). In the first step, authors one and two drew on existing sources in order to develop an initial coding scheme. One source we drew from was Brennan and Resnick (2012) discussion of computational thinking practices (including being incremental and iterative, testing and debugging), and computational thinking perspectives (personal expression, connecting with others, and questioning one's environment). We also drew from our prior analysis on portfolios from year 2 of the project (Lui et al., 2018; Shaw et al., 2019), where we considered what students were most likely to report upon in terms of their process and learning, especially regarding collaboration, aesthetics, and personal learning.

The two researchers applied to these initially generated coding categories to 15 randomly chosen portfolios from the three classrooms where teachers had prior experience with e-textiles (5.9% of students). In particular, we focused on the two latter sections of the portfolio (*Process* and *Reflection*), which emphasized students' experiences constructing their projects, and their self-reported descriptions of what they learned. We then wrote analytical memos (Saldana, 2011) about this initial application of these categories, which we used to further refine categories until we created our final coding scheme (see online supplemental materials), for instance, clarifying what subcategories should fall under the larger category "Design Changes". We reapplied the final new scheme to the 15 originally selected portfolios until we established 100% consensus between the first two coders. We finally trained an additional three coders (Author4, Author5, and Author6) on the new coding scheme using the same set of portfolios, in preparation for coding all the available portfolios.

In order to reach inter-rater reliability, the five coders then jointly coded a randomly selected set of portfolios (9.8% – 20 of 221 portfolios across all classrooms) where we established strong inter-rater agreement with an intraclass correlation coefficient = 0.75; $p < .001$ (a correlation of 0.81 and above is nearly perfect and above 0.70 is considered acceptable). Among the five researchers, we divided and coded the remaining portfolios (n = 201). In the analysis, we calculated the number and percent of portfolios within each of the codes described above to help judge how predominant the codes were across students' portfolios.

Finally, for RQ3 regarding a comparison between portfolios of students taught by teachers with and without prior experience with e-textiles, we compared the findings for both RQ1 (rubric scores) and RQ2 (coding for computational practices) between classes taught by novice e-textiles teachers (12 teachers, teaching e-textiles for the first time) versus classes taught by experienced e-textile teachers (3 teachers, teaching e-textiles for the second or third time).

Findings

Our findings are divided into two sections. First, we report on the quality of student communication about computation seen in the submitted portfolios using the rubric evaluations (RQ1). Second, we report on what students' shared about their computational thinking practices within the portfolios (RQ2). For both sections, we compare the findings between students who were taught by novice e-textiles teachers (teachers who were teaching the unit for the first time) versus students taught by experienced e-textile teachers (the three teachers who piloted the unit for 1–2 years previously) (RQ3).

Student communication about computation

Overall, students who submitted portfolios demonstrated a solid level of competence in communicating about their e-textile product, process, and learning. For all 15 classrooms, the scores ranged from 3 to 100 (out of 100), with a mean of 75.0, median of 75, a mode of 100, and a standard deviation of 24.2. Except for the few students who purposefully submitted incomplete portfolios (but likely prioritized finishing the projects themselves), we consider scores above 70 very high. The summary of rubric scores are shown in Table 3.

Notably, the overall portfolio rubric scores were generally higher for students who had experienced e-textiles teachers, with a range of 35 to 100, a mean of 85.9 and a standard deviation of 14.67. This tendency toward higher scores was present throughout all sections of the portfolio, for which may have occurred for several reasons. First, the experienced teachers all made enough time to complete the entire e-textile unit and allowed sufficient time for students to create their portfolios. Further, all three experienced e-textile teachers already had one year of experience implementing the portfolio as part of the curriculum, and themselves provided feedback that led to the design changes between the second and third iterations

Table 3. Average rubric scores across portfolio sections by textual and visual communication.

	Textual Description Score Average			Use of Visual Evidence Score Average		
Portfolio Section	All portfolios	By students with *experienced* teachers	By students with *inexperienced* teachers	All portfolios	By students with *experienced* teachers	By students with *inexperienced* teachers
Project		Description	84.2%	91.0%	82.8%	74.6%
88.7%	71.7%					
Process 87.8%	49.2%	Description	75.4%	91.9%	72.1%	55.6%
Reflection on Learning	79.0%	90.1%	76.8%	52.4%	78.9%	47.0%
OVERALL	78.5%	91.2%	75.9%	59.5%	85.8%	54.2%

of the portfolio (reported in the background section of this paper and in Lui et al., 2018). The comparison between the rubric scores between experienced, novice, and all e-textiles teachers is detailed in Table 3.

Textual computational communication

In terms of the students' textual communication performance, students generally scored the highest within the *Product* section (mean 84.2%), where they were asked to describe their final project. This may have occurred because this section required factual, concrete descriptions of the structural and functional aspects of students' projects, including where the LEDs and sensors were placed, as well as what their light patterns did (e.g., LEDs blinking back and forth slowly). This can be seen in Fernando's description of his project illustrates a strong textual project description[2]:

> In this project what we did was that we made a 3D figure of something that could've been a pillow, teddy bear backpack, etc. what i went with was a pillow, the pillow had to have three LED lights and two patches that when touched with a specific amount of pressure the pattern of the lights would change. On the pillow i had a dinosaur figure and the 3 lights were placed in a row on the tail, the patches were placed to the side. I had 4 patterns of the lights

1.Pattern 1 show us 3 lights taking turns turning on and off
2.Pattern 2 each light rapidly blinks at its own time
3. Pattern 3 two lights turned on at the same time while the other was off then they switched
4.Pattern 4 all lights slowly turned on in a sequence

Here, Fernando precisely outlined the details of both his physical project form and light patterns that he coded, even describing the choice he made in creating a pillow instead of a different form of project. Each lighting pattern is clearly described within a numbered list. Because of his inclusion of all the required details, as well his clarity of description, Fernando received full credit for this section of the rubric.

Students scored slightly lower in textual communication in the remaining two sections of the portfolio: *Process* (mean 75.4%) where students had to describe their challenges faced and/or revisions made, and *Reflection* (mean 79.0%), where they described their growth and learning. There are several reasons why this may have occurred. First, students may simply struggle more with describing processes more than products. Describing a process moment requires thinking across time and being aware of the reasons for one's actions. Second, process moments took place while projects were being made and often *before* the creation of the portfolio itself; unless students had documentation (such as design notebook entries) about mistakes or revisions they made, it may have been more difficult for them to remember details. Third, many students who provided high levels of detail occasionally lost points because they focused only on *one* process moment instead of the required *two*. This makes sense considering the fact that even describing a single moment by providing complete details and clarification required more effort and space than vaguely describing two moments.

Below are two examples of students' textual descriptions of process moments that illustrate the range of detail that students included. Here, both students wrote about dealing with sewing/crafting and related circuitry issues in their projects. While both students use the span of a single paragraph, Alma described two process moments with vague and general language. Dhruv, on the other hand, described one sewing issue in detail within a single paragraph:

> My first obstacle was the outline of my project. My problem was that they [conductive thread lines] crossed with one another. I tried a couple of times (which wasn't a success),then tried a different setup and worked perfectly. My next challenge was my sewing. The first time I sowed I had messed up pretty badly. Only 3 of my lights were turning on sol decided to re-sow. Which then ended up workin (Alma, p. 4)

When I had begun sewing my final project, I faced an obstacle. Both the foils [touch sensors], negative and positive, were not reading the amounts of pressure applied to the project to change the light patterns. I tried to redo the foil to create more connectivity from the foil to the conductive thread. I redid the foil and chose a much bigger piece this time. Instead of sewing all over the foil, I decided to go around the whole foil but around on the edges. This made a huge difference, and I was receiving much better results from the foil immediately. The project is now able to read all the four different light patterns depending on the amount of pressure applied on the foils. (Dhruv, p. 3)

In the above descriptions, both students reported on their problems and subsequent solutions, though to differing degrees of detail. While Alma describes two problems in this excerpt, both descriptions lack specific details about the actual nature of the problem (e.g., "The first time I sowed I had messed up pretty badly. Only 3 of my lights were turning on") and its solution (e.g., "I decided to re-sow. Which then ended up workin"). Dhruv, on the other hand, provided not only more details about what his problems were ("Both the foils [touch sensors], negative and positive, were not reading the amounts of pressure applied to the project to change the light patterns"), but also outlined two major changes he made to resolve it: larger foil patches and better sewn connectivity across a greater area of the patches. Notably, Dhruv's second challenge report (not included here, but also a requirement of the portfolio) was just as highly detailed as the paragraph above. Some students received less credit for their process moments because they only included one moment, even if it was described in great detail. This reveals some of the limitations of the portfolio guidelines and rubric, which did not provide as much information on the level of detail in students' textual explanations so long as they included an issue and the resolution.

In the *Reflection* section, many students were able to communicate textually about the concrete hard skills and competencies they acquired (e.g., sewing, programming). However, these descriptions were generally less clear and detailed when writing about abstract ideas such as improved soft skills (e.g., creativity, teamwork, planning). While students mentioned their soft skills, it was often difficult to judge whether or not they could provide the requisite evidence for the development of these skills. For instance, David chose to describe improvement in sewing skills, which he demonstrated with three pictures and explained textually:

> Through the E-textiles unit, I have improved on one skill drastically, and that skill is sewing ... [proceeds to show three images that highlight growth in sewing ability across different projects] As you can see, my sewing skills improved drastically throughout these projects.

> They went from a huge mess to clean stitches with cleaner cut offs. What I didn't take into consideration in the beginning that I think made my sewing look bad is which side I would [tie and] cut the sewing off. On the first project, I cut the thread off on the front, so that made it look bad. (David, p. 5)

Here, we can see how David was able to include concrete details ("clean stitches with cleaner cut offs ... " and knowing "which side I would [tie and] cut the sewing off") to demonstrate his growing ability to sew an e-textiles project. Another student, Klinge, had more difficulty describing how his design skills had improved:

> One thing I improved on in the e-textiles is designing. I learned how to make design and follow through with them. Designing really helped me get to my final product. It was difficult designing it and choosing which port the conductive thread would go to. (Klinge, p. 7)

Klinge describes that he improved on "designing" but provides little detail about how that design actually changed besides the fact that he was able "follow through with them". While he mentions the act of figuring out the circuit connections of the project ("choosing which port the conductive thread would go to"), he does not detail whether or not this is an aspect of the design, or through what means this activity became easier for him (e.g., changing the spatial arrangement of components, deciding upon the grounding line for components first).

This challenge of providing details for less concrete skills speaks to the need to further scaffold students when reflecting on developing abstract or soft skills within future portfolio designs. Students may need help defining what an improvement in collaboration, design, planning, and similar soft skills looks like in detail. Further, as we discuss below, these abstract soft skills proved particularly difficult when thinking about visual communication as well.

Visual computational communication

Students who submitted portfolios demonstrated a basic level of competency with visual communication, albeit with a much lower score than on textual communication (mean: 59.5%). As with textual descriptions, students overall scored much higher in the *Product* section (mean 74.6%). As with the textual communication, this is likely due to the concrete nature of reporting on the final project itself, and to the fact that images could easily be gathered at the last minute to support their descriptions. Students here featured final pictures of their projects, and many added annotations such as arrows or captions to help explain the placement of components. As an example, Kristin provided two images in this section, which included an overall view large enough to see the placement of the patches on a plushie doll as well as a close-up figure of the circuitry of the lights in the doll's heart (See Figure 2). Note that her visuals are carefully annotated with arrows and labels pointing to the lights and an explanation for why she needed two diagrams.

Within the *Process* and *Reflection* section, students scored much lower for visual communication (55.6% and 52.4% means, respectively). Here, some students did not annotate or label their evidence, while others provided no visuals at all. As with the textual communication, one potential reason why students received lower scores in the process section was because of timing issues. When dealing with issues during construction, many students were likely more involved in actually making their projects work, rather than stopping to visually document their mistakes, such as with screenshots of

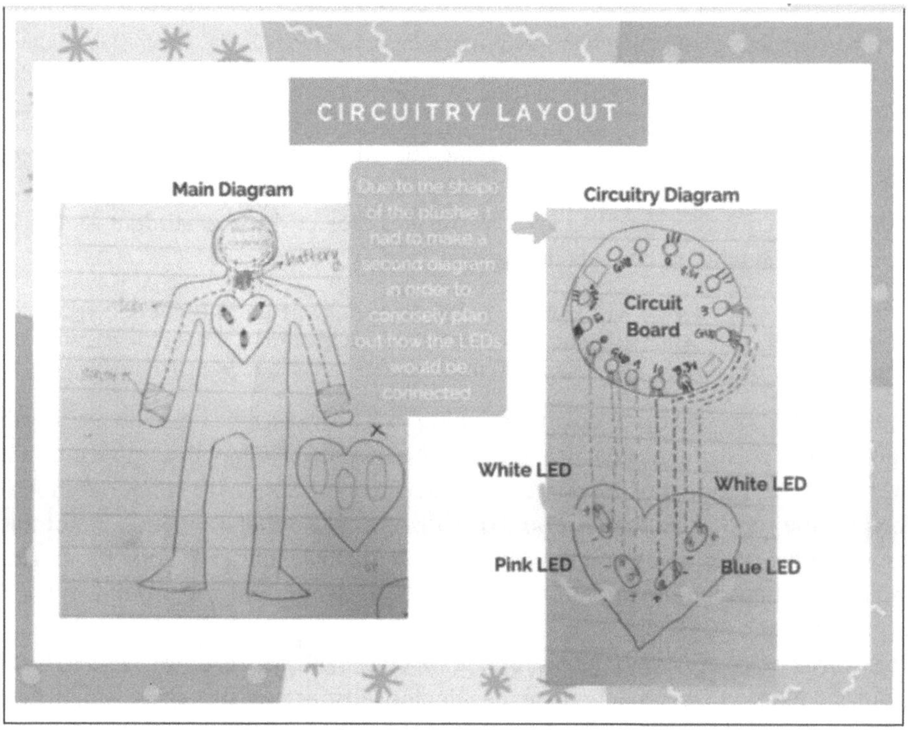

Figure 2. Kristin's visual explanation of the design and circuitry of her human sensor project (p. 3).

errors in their code, or pictures of messy sewing that caused short circuits. Notably, this lack of recorded evidence is something that some students explicitly mentioned within their portfolio (e.g., "No pictures of this").

To demonstrate the range of annotation that students provided for visual evidence, we share three examples. Nino's explanation of fixing an error in his code demonstrated what effectively annotated visual evidence can look like (see Figure 3). Alongside a substantial textual explanation of the change he made in coding the ranges of his sensor conditions, Nino provided before and after screenshots of his code with detailed captions that explain exactly what changes he made. Using a different but similarly effectively method, Angelica provides an image annotated with arrows and labels (alongside a detailed textual description) to compare changes she made within her circuit diagram (see Figure 4). In contrast, Cesar also provided multiple images to accompany his process moment description, however, he does not annotate or caption these to aid in supporting his descriptions. Here, it is unclear how these pictures relate to the two challenges he describes ("Figuring out the coding" and "getting the wires to fit on the bat") since they do not show his excerpts of his code, nor highlight the stitching on the project (see Figure 5).

The vast range of performance of students on visual annotation of process moments illustrates the need to further support students with documenting (as well as solving) these moments of troubleshooting and debugging. The differences between students who were taught by an experienced versus novice e-textile teachers is particularly pronounced in this area (87.8% versus 49.2%, respectively). This suggests that the experienced teachers may

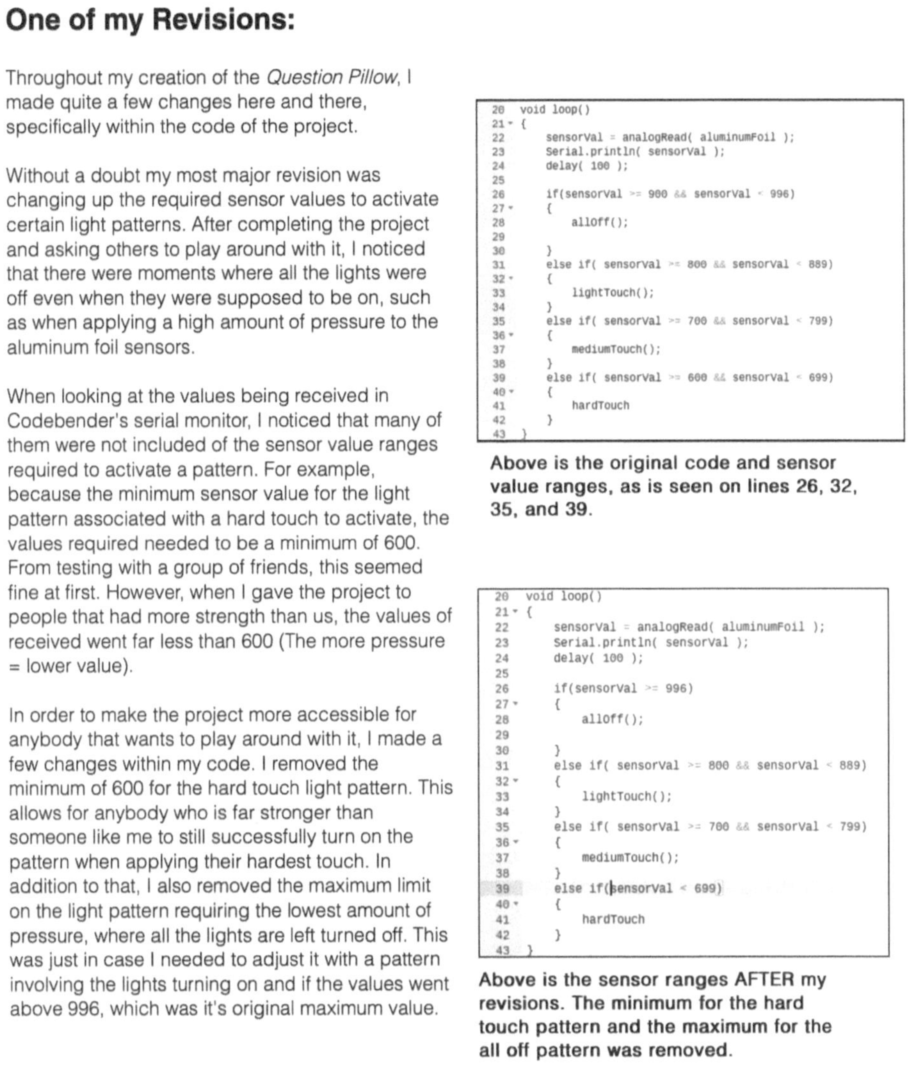

One of my Revisions:

Throughout my creation of the *Question Pillow*, I made quite a few changes here and there, specifically within the code of the project.

Without a doubt my most major revision was changing up the required sensor values to activate certain light patterns. After completing the project and asking others to play around with it, I noticed that there were moments where all the lights were off even when they were supposed to be on, such as when applying a high amount of pressure to the aluminum foil sensors.

When looking at the values being received in Codebender's serial monitor, I noticed that many of them were not included of the sensor value ranges required to activate a pattern. For example, because the minimum sensor value for the light pattern associated with a hard touch to activate, the values required needed to be a minimum of 600. From testing with a group of friends, this seemed fine at first. However, when I gave the project to people that had more strength than us, the values of received went far less than 600 (The more pressure = lower value).

In order to make the project more accessible for anybody that wants to play around with it, I made a few changes within my code. I removed the minimum of 600 for the hard touch light pattern. This allows for anybody who is far stronger than someone like me to still successfully turn on the pattern when applying their hardest touch. In addition to that, I also removed the maximum limit on the light pattern requiring the lowest amount of pressure, where all the lights are left turned off. This was just in case I needed to adjust it with a pattern involving the lights turning on and if the values went above 996, which was it's original maximum value.

```
20  void loop()
21  {
22      sensorVal = analogRead( aluminumFoil );
23      Serial.println( sensorVal );
24      delay( 100 );
25
26      if(sensorVal >= 900 && sensorVal < 996)
27      {
28          allOff();
29
30      }
31      else if( sensorVal >= 800 && sensorVal < 889)
32      {
33          lightTouch();
34      }
35      else if( sensorVal >= 700 && sensorVal < 799)
36      {
37          mediumTouch();
38      }
39      else if( sensorVal >= 600 && sensorVal < 699)
40      {
41          hardTouch
42      }
43  }
```

Above is the original code and sensor value ranges, as is seen on lines 26, 32, 35, and 39.

```
20  void loop()
21  {
22      sensorVal = analogRead( aluminumFoil );
23      Serial.println( sensorVal );
24      delay( 100 );
25
26      if(sensorVal >= 996)
27      {
28          allOff();
29
30      }
31      else if( sensorVal >= 800 && sensorVal < 889)
32      {
33          lightTouch();
34      }
35      else if( sensorVal >= 700 && sensorVal < 799)
36      {
37          mediumTouch();
38      }
39      else if(sensorVal < 699)
40      {
41          hardTouch
42      }
43  }
```

Above is the sensor ranges AFTER my revisions. The minimum for the hard touch pattern and the maximum for the all off pattern was removed.

Figure 3. Nino's labeling of code screenshots before and after fixing his sensor ranges (p. 4).

have helped their students better prepare for reporting on these process moments better during the weeks of project creation including the suggested scaffolds mentioned in the methods section (e.g., in-class journal prompts, use of exemplars, class conversations).

Generally, students also scored lower for visual communication in the *Reflection* section. This parallels trends within textual communication for this part of the portfolio, especially when thinking about students' reporting on hard versus soft skills. For example, consider John's description of how his sewing improved (see Figure 6) where he compared the sewing of his first project (in yellow, on left) to the sewing on his final project (in black, on right). This image accompanied a textual explanation describing how he learned to sew smaller stitches that conducted electricity better (because the thread is more taut). Thus, students who reported on hard skills (e.g., sewing, drawing) could more easily present visual evidence of their improvement because it was more concrete. Students

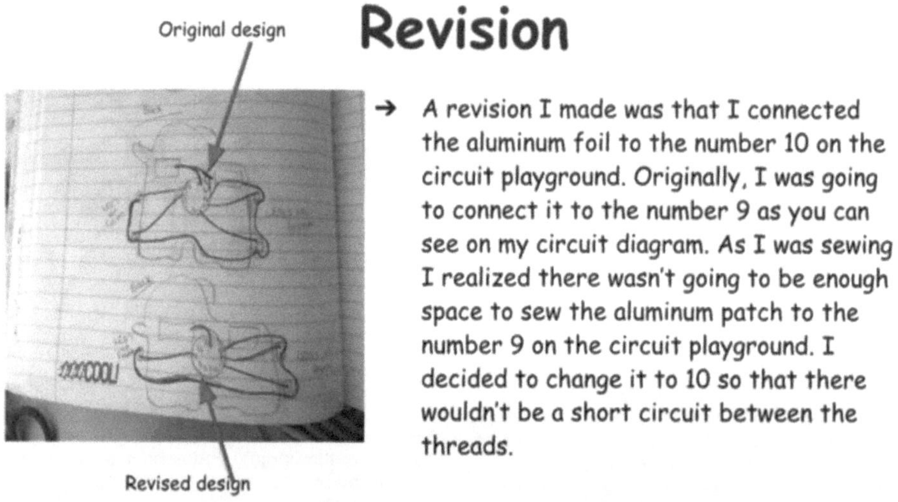

Figure 4. Angelica's annotation of changes in her circuitry diagram (minute 1 in portfolio video).

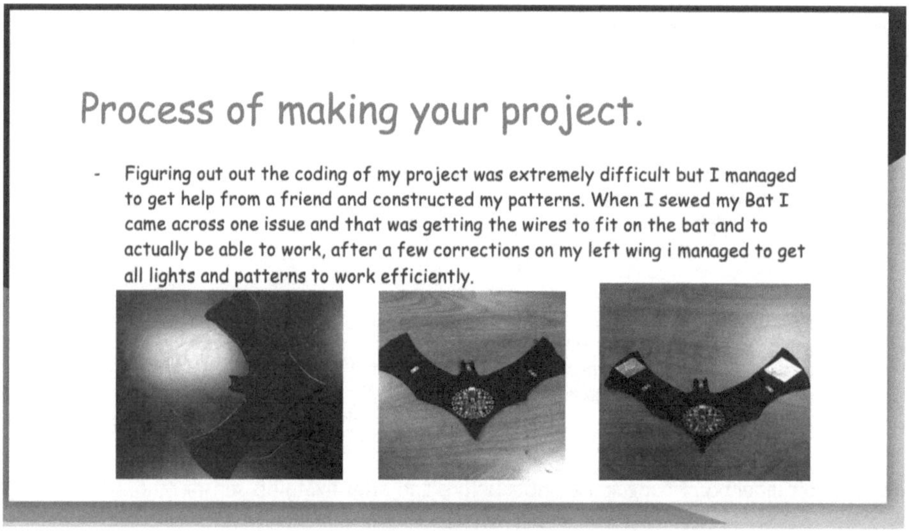

Figure 5. Cesar included visuals with no annotation or textual explanation for them (p. 3).

who reported on soft, abstract skills (e.g., planning, patience) tended to present less images in general to support their narrative. This difference can be seen in Zoe's portfolio, where she explained:

> Sure I learned how to sew, but what's most important to me that I know I learned is patience. Patience was a huge key involving this unit, not only in computer science, but patience has helped me throughout many classes this year. And so, without patience, you cannot really accomplish anything in life. (p. 6)

Here, Zoe does not provide any images to support her growth in patience, and even comments that she could have focused on sewing as an area of growth (which might

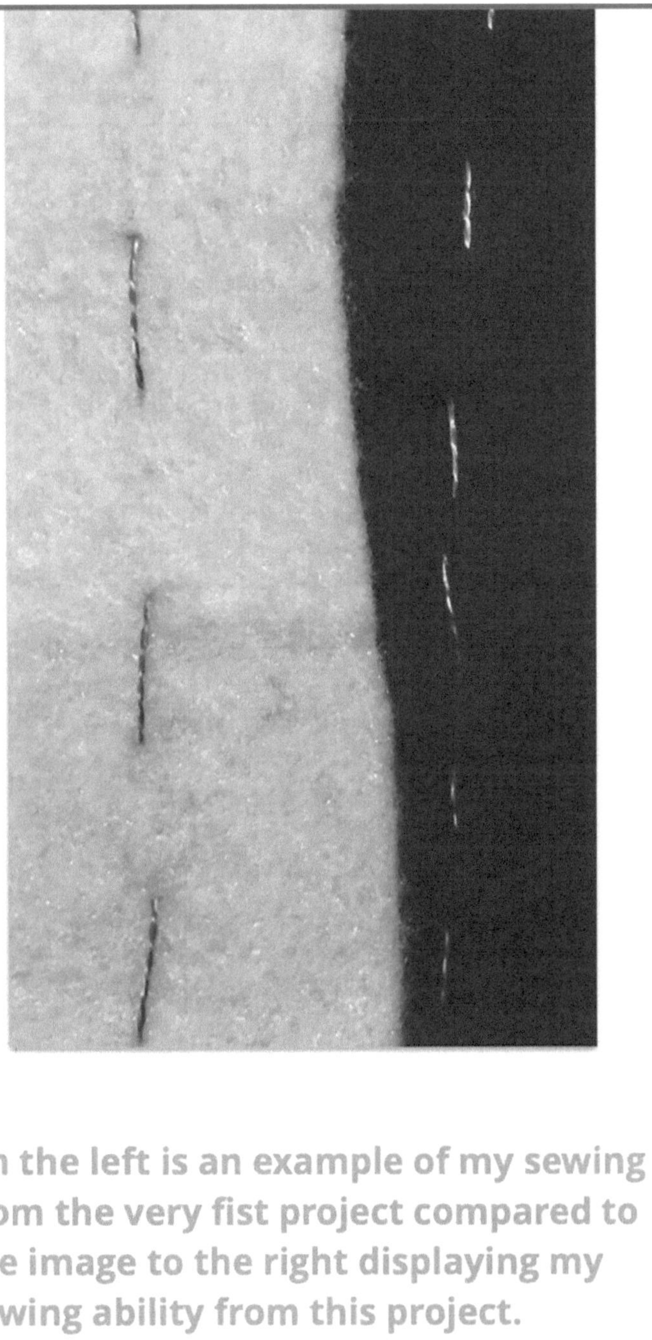

Figure 6. John's picture and annotation comparing sewing in his first and last projects, demonstrating growth through smaller stitches.

have better lent itself to visual documentation) but chose not to. It is difficult to imagine what kind of imagery could evidence her growth in patience here, an area that she claims has influence on other scholastic areas of her life. This speaks to the need for us to reconsider how either form of communication (textual or visual) might not always be the best way of communicating particular types of computational information, and how to appropriately support students in choosing the most effective forms for sharing their ideas.

Students' reports of computational thinking practices

When reporting on their process of creating their e-textiles, students were required to describe either the design revisions they made or the problems they encountered along the way. While the requirements for these sections were simple (i.e., including problems *and* solutions, or revisions *and* justifications), some students' more detailed descriptions referenced a number of rich computational practices. Below, we explore the computational practices that students chose to report on, including: 1) debugging and problem solving, 2) iteration and revision, 3) collaborating, and 4) using external resources and planning. Here we are less interested in the predominance of students who reported these computational practices (though we detail the percentages) than the actual presence of these descriptions in student-created portfolios, as they show the potential for process-based portfolios to document these under-assessed computational practices in students' open-ended project design.

Debugging and problem solving

Within the *Process* section of portfolio, most students chose to explain at least one problem that arose while creating their e-textile designs (contained within 84.6% of all portfolios) (see Table 4 for subcategories). Almost all students who reported on a problem were able to successfully describe how they first *observed this issue* (83.7% of all portfolios), while a majority were also able to *describe the specific solution* they implemented (69.2%). While describing both a problem and its matching solution may seem like simple tasks, being able to identify and describe a particular issue within the context of a larger, complicated project is both a challenge and a recognized aspect of computational practices such as analyzing, decomposing, and debugging problems (e.g., Brennan & Resnick, 2012; College Board, 2017; Grover & Pea, 2018). The range of problems described in the portfolios were situated within all the

Table 4. Computational practices students described in problem-based process moments.

Problems Discussed			
Practice Described (not mutually exclusive)	All portfolios	By students with experienced teachers	By students with inexperienced teachers
Observing an Issue/Problems	83.7%	93.8%	79.6%
Describing a Solution	69.2%	82.8%	63.7%
Testing and/or Refining Something Iteratively	19.5%	31.3%	14.7%
Isolating a Problem/Root Cause	9.5%	15.6%	7.0%
Other	1.8%	3.1%	1.3%
At least one practice	84.6%	95.3%	80.25%

domains of e-textiles (coding, crafting, and circuitry), and often included overlaps in those domains, something that frequently occurs during the e-textiles construction process (Kafai et al., 2014).

To understand the type of thinking that went into documenting these problems, consider Valentina's account of dealing with syntax problems which included an explanation of how these errors negatively impacted the functionality of her program:

> One challenge in the coding was forgetting brackets and semicolons. First, when we initially tested the code, one light blinked and nothing else happened. While troubleshooting, we discovered that a few brackets were in the wrong place (causing the delays to loop) and a few semicolons were absent. To fix the issue, we checked each line for brackets and semicolons and our code carried out the desired function and was successful! (p 4)

In this excerpt, we see how Valentina is able to reflect on her process of debugging by breaking it down into multiple steps. This includes: 1) discovering the issue ("one light blinked and nothing else happened"), 2) diagnosing the cause ("a few brackets were in the wrong place … and a few semicolons were absent"), and 3) implementing a solution (systematically "checked each line for brackets and semicolons"). Further, we can see evidence of her strong comprehension of the code, where she successfully identifies how syntax errors can change the logical output of her program (i.e., how misplaced brackets were "causing the delays to loop").

In the course of reporting on problem solving and debugging, a number of students also demonstrated engagement in the computational practice of *isolating a problem or root cause from multiple possibilities* (identified in 9.5% of all portfolios). This can be seen in the portfolio of Luz, who wrote about how she was able to diagnose the cause of her dysfunctional LEDs: "I hooked the [microcontroller] board up to a seperate LED using alligator clips and it worked, so I knew that the issue was with the wiring on the inside" (p. 3). Here, Luz describes how she first had to check the functionality of her program by "hook[ing] up her board to a seperate LED", thereby isolating her written code from her sewn circuitry. After realizing that the code was indeed functioning properly, she then was able to identify that "the issue was with the wiring on the inside." Finally, she was able to fix the stitched wiring so that the lights worked. Thus, even with this short quote, we can see Luz' complicated process of moving across the three domains of e-textiles: from programming, to circuitry, to crafting.

Another debugging/computational practice that students were able to illustrate in their portfolios was *testing and/or refining something iteratively* (19.5% of all portfolios). Nino, for instance, (see Figure 3 above), wrote about the process of repeatedly adjusting the sensor reading ranges within his code to accommodate a range of users who would be touching the aluminum foil sensors in his interactive artifact:

> When looking at the [sensor] values being received in Codebender's serial monitor, I noticed that many of them were not included of the sensor value ranges required to activate a pattern … From testing with a group of friends, this seemed fine at first. However, when I gave the project to people that had more strength than us, the values of received went far less than 600 (The more pressure = lower value).

In order to make the project more accessible for anybody that wants to play around with it, I made a few changes within my code. I removed the minimum of 600 for the hard touch light pattern … [which] allows for anybody who is far stronger … to still successfully turn on the pattern when applying their hardest touch … I also removed the maximum limit on the light pattern requiring the lowest amount of pressure … (p. 4)

Here, Nino demonstrates his ability to connect the multiple rounds of testing his project (both with "a group of friends" and "people that had more strength") to the actual changes that he made to his code (changing the minimum and maximum pressure sensor readings written into the logical expressions of his conditional statements). He explicitly mentions how these changes are meant to make his "project more accessible" to a wider range of users. From this perspective, students' communication of their processes make this practice of testing and refinement more visible to teachers and researchers, something that would normally be hidden when only evaluating a students' final products.

Finally, a few students mentioned other miscellaneous problem solving practices (1.8% of all portfolios), including collaborative checking of problems, or learning from other people's mistakes and solutions. Overall, students in classes with experienced e-textile teachers tended to show increased percentages of describing solutions, isolating problems, iterative testing/refinement, or other problem solving practices (see Table 4). As highlighted in our previous work (Lui et al., 2018), one major reason that might explain this difference is that these more experienced teachers had previous experience implementing the portfolios within the e-textiles curriculum. Based on this, they made changes in their classrooms to more explicitly promote high levels of computational communication in their classroom, leaving more time at the end of the unit for students to focus on creating their portfolios, as well as giving them greater opportunities to reflect and record their progress using the suggested journal prompts. Beyond these differences, however, students' high levels of discussion of their debugging practice, as well as some students' description of more complex computational practices such as isolating problems and testing/iteration illustrates the affordance of portfolios in not only sharing these experiences with their teachers, but also allowing students to externalize their thinking in the concrete format of the portfolio.

Iteration and revision

Students' descriptions of their project revisions (40.3% of all portfolios) created additional opportunities to examine their engagement with computational practices. In our coding scheme, we categorized problem solving as the act of fixing some error during the process of construction, while revisions were categorized as changing the underlying design of a project for a wide range of reasons that did not necessarily relate to functionality. The most widely cited reason for changing a design related to spatial composition (22.2% of all portfolios), or the arrangement of components and/or circuit lines on the physical surface of the project. Within the portfolios, many students mentioned how spatial composition became difficult when dealing with the three-dimensionality of projects. One student, Seth, demonstrated this when we described adjusting his design for his "stuffed taco" project and dealing with both inside and outside surfaces of the project. While he sewed his microcontroller to the inside of his toy, he did not realize that this would cause issues when connecting an external battery to the project. Changes he described included adding a little hole into his fabric to make plugging in the battery easier, as well as adding more "fluff" to the inside his project in order to isolate the stitches and prevent potential short circuits.

Other reasons that students provided for revisions included aesthetic and audience-related justification (11.8% and 5.0% of all portfolios, respectively). Joshua, for instance, reported on aesthetic changes he made to shape of his sensors:

> A revision that I made on my project was in my design. At first the foil sensors were the shape of a square.But then I decided to change them to mustache's so that they would look similar to the main design which was the mustache. (p. 4)

We can see that Joshua made a decision to change a functional element of his project (the square-shaped foil sensors) just for the purpose of matching his overall project design (a fabric mustache). Thus, while sometimes aesthetic justifications related more to personal expression (e.g., disliking how something looked), other times these reasons were tied to creating a more coherent design. Fewer portfolios (5.0% of all portfolios) mentioned usability or audience need-reasons for revisions, which included moving human sensor patches to an area which made them easier to handle, or revising the location of felt and/or LEDs to make a character and/or pop culture icon easier to recognize. While engaging with these aesthetic or usability concerns may not fit within existing lists of computational practices, they are legitimate justifications communicating the "design and appropriateness" of student choices (College Board, 2017).

Finally, numerous students provided other reasons for revisions that did not fall into any of the categories above (7.2% of all portfolios). While these justifications varied, a number of students described thinking about work efficiency, that is, redesigning a project to make it easier to construct. Mostly, this included reorganizing circuit lines to save thread, sewing effort, or actively prevent future dysfunctions such as short circuits (as seen with Seth above). This attention to efficiency of design is reminiscent of programmers' valuing of efficient code – doing more with fewer lines – but related more to physical craft than to code.

When comparing the revisions described by students with experienced versus noviced e-textile teachers, we see similar trends to both textual and visual communication as well as reporting practices of problem-solving (see Table 5 below). Slightly more students who had experienced e-textile teachers reported on spatial and aesthetic justifications, while fewer reported on "other" or listed no reasons for their changes. By far, the greatest difference occurred within the usability/audience category, with 14.1% of students with experienced teachers reporting on these reasons versus 1.3% of students of novice teachers. This difference can be attributed to a particular teaching practice that one experienced teacher adopted: a mini-lesson where he gave his students opportunities to test their projects with their classmates (as described with Nino above, Figure 3). Not

Table 5. Computational practices students described in revision-based process moments.

Revisions Discussed			
Reasons for revision (not mutually exclusive)	All portfolios	By students with experienced teachers	By students with inexperienced teachers
Spatial	22.2%	34.4%	17.2%
Aesthetics/Expression	11.8%	20.3%	8.3%
Usability/Audience	5.0%	14.1%	1.3%
Other Reason	7.2%	4.7%	8.3%
Reason not provided	4.5%	1.6%	5.7%
At least one reason	40.3%	57.8%	33.1%

surprisingly, many students in this class reported on these tests and subsequent revisions of their human sensor reading ranges in an attempt to accommodate different users.

In sum, all the revisions and related justifications described above illustrated the students' iterative improvements on their projects, a key computational thinking practice. Additionally, discussion of these revisions also illustrates students' engagement with other important, yet under assessed CT practices, whether creative expression through aesthetic decision making, attentiveness to usability and audience, or thinking about efficiency of computational design and work.

Collaborative work

Our open-coding also revealed students' engagement with the computational practice of collaboration (see Table 6). Over one-third of the portfolios included descriptions of working with other people (35.8% of all portfolios), whether teachers, classmates, friends or family. Most students reported collaborations with their classmates (26.7%), with fewer writing about teachers (9.1%), and even fewer about unspecified or other people such as family (3.6%). As described in the sections prior, this can be seen within reports of working with peers on testing projects. Additionally, numerous students also mentioned actively receiving help from others, as Lili described: "I asked for help with my code from my fellow peer and teacher multiple times." (p. 4). Reporting on these activities is notable since we did not explicitly ask for students to comment upon collaboration, and the portfolios were predominantly an individually graded project. Additionally, while CS is traditionally spoken about in individual terms (Yardi & Bruckman, 2007), this illustrates how e-textiles production can become a social practice, especially when considering the highly visible and tangible nature of the projects themselves (see Fields, Jayathirtha et al., 2019).

Planning and using external resources

Students also wrote about drawing from the wider environment while creating their e-textiles projects through use of external (non-human) resources, materials, and tools (8.6% of all portfolios). For instance, Serena wrote about referring "back on past [e-textiles] projects to help me find solutions" (p. 5), while Alicia described remixing existing sample code for fading LEDs, which she found in the Arduino IDE, for her own project. These reports therefore illustrated how students' individual engagement in creation were still situated into the larger context of the unit and classroom (e.g., referencing past e-textiles projects) and in the broader DIY and open source communities of e-textiles (e.g., referencing sample code from the Arduino IDE).

Table 6. Types of collaborators students described.

Collaboration			
Collaborators (not mutually exclusive)	All portfolios	By students with experienced teachers	By students with inexperienced teachers
Classmate	26.7%	28.1%	26.1%
Teacher	9.1%	6.3%	10.2%
Other (family; neighbor)	3.6%	6.3%	2.6%
At least one collaborator	35.8%	37.5%	35.0%

Another practice that students mentioned within their process descriptions, but which was not explicitly asked for, was planning (27.6% of all portfolios). Discussions of planning included students' explicit attempts to avoid problems and mistakes in the future, as well as organizing workflow for project completion. This can be seen within an excerpt from Dana's portfolio, where she described her plan for sewing:

> Sewing was pretty hard. I had to stitch from the back of the shirt to the front. Since the LED lights were in the back I had to make sure I had enough [thread] to make it around. The [thread] would sometimes get mixed up or get tangled. (p. 3)

Here, Dana's description of her workflow ("stitch from back ... to the front"), and its relation to her materials (ensuring she "had enough [thread] to make it around") and potential issues to monitor ("mixed up or ... tangled" thread) illustrates her thoughtfulness when thinking about issues of construction.

Relatedly, a number of students also mentioned the process of idea generation (8.6% of all portfolios) as part of their computational processes. Often mentioned as a problem to be overcome, students mentioned their "solutions" including drawing and sketching, looking through different objects or accessories at home, or talking through project ideas with others. As seen in Table 7, students with experienced e-textile teachers tended to report on planning and idea generation more frequently, again following trends throughout portfolio descriptions of more effective communication strategies, as well as sharing more information about their project-creation experiences.

Discussion

In this paper, we examined the assessment of the often ignored practice of communicating about computation. Just as students need to learn computational concepts and practices required for creating and debugging programs, they also need to be able to communicate about their products and processes. This is important not only for working with teachers and classmates, but any future collaborators down the line. Communication about computation can take various forms, from casual conversations with peers about troubleshooting their projects, to more formal documentation of code presented to users and clients. In our case, we chose portfolios as a multi-modal format to capture students' communication about their e-textile products and processes.

Our findings reveal that students who created process-based portfolios documenting their e-textiles projects generally illustrated a solid ability to write and visualize computational ideas within their portfolios, albeit with room for improvement in several areas. The portfolios also allowed for students to report under-assessed computational practices

Table 7. Other computational practices described in Portfolios.

Other Computational Practices			
Categories (not mutually exclusive)	All portfolios	By students with experienced teachers	By students with non-experienced teachers
Using External Resources (sample code; alligator clips; class notes; previous projects, etc.)	8.6%	11.0%	7.6%
Planning	27.6%	43.8%	21.0%
Coming up with Ideas	8.6%	20.3%	3.8%
At least one other practice	38.9%	59.4%	30.6%

such as debugging, iterating, and collaborating. Notably, the quality of students' communication about their projects and processes in classrooms with experienced e-textile teachers (one or more years of e-textiles) was substantially better than those with novice e-textile teachers (teaching e-textiles for the first time). This difference in performance indicates how important both teacher preparation and classroom implementation are in supporting these communication skills. In the following sections, we (1) discuss what we learned about students' abilities to communicate about computation, and (2) address challenges of implementing portfolios in classrooms.

Assessing computational communication

Our study shows that portfolios using both text and visuals can be used as an insightful form of assessment, especially for open-ended projects (not necessarily just e-textiles). Not only does the creation of a portfolio give students opportunities to rehearse this important skill, but also allows researchers and practitioners to capture aspects of their computational practice not easily captured by other assessments, such as their approaches toward debugging or their process of revisions. This knowledge of student learning would be difficult to gather through other means outside of time-consuming reflective interviews or think aloud problem solving sessions (e.g., Grover, 2017; Jayathirtha et al., 2020). For this reason, portfolios should be seen as a viable part of a larger "system of assessments" (Grover, 2017) of students' computational thinking. In the e-textile unit, this system of assessments included small tasks such as design notebooks, journal entries, and short debugging tasks, as well as larger tasks, namely the set of four projects (each with a rubric for grading) and the portfolios.

In terms of the students' communication ability, our rubric analysis shows that students who submitted portfolios possessed a solid aptitude in communicating about their computational projects and processes. As described in our literature review, students' abilities to communicate about computation can be measured through three levels of competence (Taffe, 1989). The first level of competence describes students' abilities to comfortably use the terminology/vocabulary of computing, while the second level describes their ability to apply this vocabulary to describing larger computational systems or projects. In our study, students demonstrated mastery of these first and second levels of communication, evidenced by the relatively high scores within the *Product* section of the portfolios, where students described the components and functionality of their own e-textiles projects. Here, students illustrated both their ability to communicate in written text and through visualizations, where we saw students using a variety of visual annotations (e.g., arrows; highlighted sections), as well as image formats (e.g., overall shots; close-ups) to support the description of their projects.

However, students still struggled with the third level of communicative competence (Taffe, 1989) which looks at how well students can marshall their computational descriptions to either support an argument, or provide evidence of a specific idea. In particular, we can see this in students' varied abilities to describe their processes and reflect on what they learned (sections 2 and 3 of the portfolios, respectively). While some students were successful in detailing their experiences in these sections, others had more difficulty when attempting to write about more abstract ideas or procedures. For example, in the *Reflection* section of the portfolio, students tended to have an easier time providing support for the hard skills they had learned (e.g., sewing, circuit design) in the form of pictures or specific details. However, discussion of soft skills learned (e.g., planning, design), though present, were more vague and lacking in evidence.

Another place where students' struggled to communicate effectively was in describing their computational processes. Our findings highlighted the abilities of *some* students to describe less concrete methods such as testing and refining something iteratively (19.5% of all portfolios), and isolating problem causes during debugging (9.5% of all portfolios). However, not all students were able to describe their processes at this level of detail, nor include relevant images (whether in-process pictures or relevant code excerpts) to support this discussion. While we did provide supports and scaffolds such as in-class journal prompts and an exemplar portfolios, this finding supports the idea that even more scaffolds to aid students in capturing and accurately describing these moments are necessary. This is something we further discuss in the section on implementation.

Despite the issues with describing process though, our analysis did uncover aspects of CT that are rarely considered in lists of computational practices, such as considerations of audience and aesthetics (e.g., Brennan & Resnick, 2012; College Board, 2017; Grover & Pea, 2018). While under-discussed, these practices are legitimate areas of program design in making computational products more cohesive or responsive. This points to the need to expand our discussion of computational thinking to include social and cultural elements, as has been called for in computational participation (Kafai & Burke, 2014). Although collaboration on a product is a recognized computational practice, creating a product to be responsive or attractive to a specific audience is a far less recognized yet valuable area of professional computing.

Drawing from these findings then, one important consideration is the design of the portfolio assignment itself. This iteration of the portfolio provided specific constraints on students in terms of how they were allowed to report on their computational ideas and processes (i.e., requiring visualizations for each section, limiting process discussion to two problems or revisions). Considering the diversity of what students reported on, however, future iterations of a computational portfolio might provide more flexibility with regard to how students decide to communicate about their own projects, processes and learning. This not only includes allowing students to decide on the scope of their own descriptions, but also what kinds of media count as evidence. For example, while we required students to submit static visualizations for each section (e.g., pictures, images, code excerpts), our analyses demonstrated how these often were not the best way of providing evidence of abstract ideas or learning. This finding gets to the larger question of not only what counts as evidence when looking to assess students' engagement with computational practices, but also considerations of how to best support students in making this determination on their own. While we focused on how students communicated *about* computation using text and visuals, future research might consider promoting communication *through* computation, such as creating a game, infographic, or animation to represent their experiences. Though beyond the scope of this study, providing flexibility within portfolios might be an avenue to promote greater agency for students, not only in terms of how they share their process with others, but also in shaping their own trajectories of computational growth and learning.

Implementing portfolio assessments in classrooms

Our findings also highlight another important item of consideration – how portfolios are actually implemented in Computer Science classrooms, and how this shapes students' abilities to communicate about computation. This relationship is clearly demonstrated through the difference in performance between students with experienced e-textiles teachers versus novice e-textiles teachers. Not only did the former group perform better throughout areas of textual and visual communication, they also tended to provide much greater details about their processes of learning, therefore shedding light on commonly under-assessed computational practices. In addition, these experienced teachers tended to have greater portfolio completion rates than novice e-textiles teachers.

As described in the methods, one major contributor to this difference in performance and portfolio completion was the scheduling and timing of the unit. Teaching e-textiles is generally a time intensive activity because it not only incorporates new programming platforms (Arduino), but also adds logistical difficulties such as dealing with and storing tangible materials and tools (e.g., sewing needles, felt, conductive thread, scissors). As detailed in our prior research, our experienced e-textiles teachers took time to re-organize their space and re-orient their teaching practice to accommodate e-textiles creation (Fields, Lui et al., 2019), something which made their third year of teaching e-textiles more efficient and productive. Almost every student from these experienced classrooms finished all four suggested projects and also completed their portfolios on time. Novice e-textiles teachers, on the other hand, struggled with timing, with some only finishing three of the four suggested projects and prioritizing finishing a functional project rather than working on the portfolio. This points to the importance of considering scheduling when trying to add portfolios to existing computing curricula. However, because experienced e-textiles teachers were able to successfully incorporate portfolios into their teaching, we believe that this challenge can be easily managed with careful planning and scheduling.

Another major contributor to the difference in performance between experienced and novice teachers' classrooms were how well they incorporated the portfolio throughout the course of the unit. While some teachers actively used the in-class supports we provided (e.g., journal prompts, exemplar portfolios), others did not. This was more prominent for experienced e-textiles teachers, who additionally made other adjustments in how they implemented the portfolios. For example, one experienced teacher ended up combining the portfolio assignment with a web design activity, such that students ended up spending more time on their portfolios than in other classrooms. Predictably, portfolios from experienced teachers' students tended to contain greater levels of detail and visual evidence.

These differences in classroom implementation and their correlation with students' communicative performance therefore suggests the importance of looking more closely at these factors in future research. Such studies need to capture not just the final days of creating portfolios, but teacher practices throughout the unit, for instance, encouraging students to pause during problems or mistakes that came up in their projects in order to both document and reflect upon their process. Another thing to note here was that this was the first process-based portfolio most students had ever created, so we might see this as a starting point for student learning rather than an endpoint. An additional line of research might consider how portfolio creation incorporated throughout the course of a unit or a school year would help students develop better communication and reflection skills. From this perspective, what is needed are observations and documentation of best practices for facilitating computational communication of design processes and problem solving with e-textiles, and in other areas of open-ended computational design.

Conclusions

We examined high school students' use of process-oriented portfolios to practice communication about computation. Overall, we found that students were capable of both writing about and visually documenting computational ideas, especially in the context of providing descriptions of their own computational designs. To a lesser degree, students were also able to report on their computational processes, demonstrating engagement with practices such as testing and/or refining something iteratively, and isolating problems and root causes. Because these portfolios were able to capture students' engagement with these under-assessed CT practices, we consider these tools a powerful way to evaluate students' computational performance, in addition to more traditional measures such as project evaluations and short-term programming tasks. That said, productive use of portfolios as an assessment tool depends not only on students' individual engagement with this task, but also teachers' ability to actively incorporate this activity into the classroom. This is supported by our findings that teachers' prior experiences and current adaptations to support portfolio creation impacted the quality of students' communication about computation both in terms of how detailed and clear the documentation was, as well as how many CT practices were described. In considering portfolios then, practitioners and researchers of CS education should additionally consider how to appropriately prepare teachers to create in-class scaffolds and ongoing opportunities for students to actively practice communication, especially in developing visualizations and reports on abstract computational processes or ideas. Only by considering all these factors can the true potential of computational portfolios for supporting communication and assessing CT practices be realized.

Notes

1. Hereafter, we refer to these three teachers as "experienced e-textile teachers" or "experienced teachers." However, we want to emphasize that all teachers in the study were experienced CS teachers; it is simply the difference in prior experience with the *e-textile unit* that we study in regard to student performance on portfolios.
2. All text from student portfolios consist of direct quotes, with no corrections of grammar, spelling or punctuation.

Acknowledgments

This work was supported by a National Science Foundation grant to Yasmin Kafai [#1509245]. Any opinions, findings, and conclusions or recommendations expressed in this paper are those of the authors and do not necessarily reflect the views of either NSF, the University of Pennsylvania, Utah State University, or University of Texas, El Paso.

Disclosure statement

No potential conflict of interest was reported by the authors.

Funding

This work was supported by the National Science Foundation [1509245].

ORCID

Deborah Fields http://orcid.org/0000-0003-1627-9512
Debora Lui http://orcid.org/0000-0003-3455-2949
Justice Walker http://orcid.org/0000-0002-4356-0396

References

Adams, T. L. (1998). Alternative assessment in elementary school mathematics. *Childhood Education, 74*(4), 220–224. https://doi.org/10.1080/00094056.1998.10521938

Ainsworth, S., Prain, V., & Tytler, R. (2011). Drawing to learn in science. *Science, 333*(6046), 1096–1097. https://doi.org/10.1126/science.1204153

Arpaci-Dusseau, A., Astrachan, O., Barnett, D., Bauer, M., Carrell, M., Dovi, R., Franke, B., Gardner, C., Gray, J., Griffin, J., Kick, R., Kuemmel, A., Morelli, R., Muralidhar, D., Osborne, R. B., & Uche, C. (2013, March). Computer science principles: Analysis of a proposed advanced placement course. In R. McCauley, T. Camp, P. Tymann, J. D. Dougherty, & K. Nagel (Eds.), *SIGCSE 13. Proceedings of the 44th ACM technical symposium on computer science education* (pp. 251–256). New York, NY: ACM.

Barron, B., Martin, C., Roberts, E., Osipovich, A., & Ross, M. (2002, January). Assisting and assessing the development of technological fluencies: Insights from a project-based approach to teaching computer science. In *Proceedings of the conference on computer support for collaborative learning: Foundations for a CSCL community*. (pp. 668–669). Boulder, CO.

Brennan, K., & Resnick, M. (2012, April). New frameworks for studying and assessing the development of computational thinking. *Presented at the 2012 annual meeting of the American educational research association*. Vancouver, Canada.

Býrgýn, O., & Adnan, B. A. (2007). The use of portfolio to assess students' performance. *Journal of Turkish Science Education*, *4*(2), 75. https://d1wqtxts1xzle7.cloudfront.net/1988152/4gf0aho874un5c6.pdf?1425082536=&response-content-disposition=inline%3B+filename%3DThe_Use_of_Portfolio_to_Assess_Students.pdf&Expires=1609450788&Signature=EjWQqjt8VDVaW5WZxZbvidGMoewffHj2-DrvCZYLalotysvbchKHimXIbbGQJ2fWi~I7UGA6-2t7b688PUP9jJ0IWYWhZ9YuanP2X88-FfkYwz5A1dgUWWzJsIsh1OEQKTfOpGqjbHZeIV42SZCTqL4LoxieCrz3QitmGBbBAgNBNuvgiRPWunx8EXQkCnWSK4cvSRA-HTc6O3ozdq660shhU-sevaGgdceETt49UZmwdW4HtB2LDZnAmzSPuo0uLm-BWYwF8IMYVeHODSwVbX05Vc1aqyxVFAgYGpwcrlwNWHLuRRFLxt38nlwvskB2lsMUjqjmQa-1gbetLZQsYQ__&Key-Pair-Id=APKAJLOHF5GGSLRBV4ZA

Chang, S., Keune, A., Peppler, K., & Regalla, L. (2015). *Open portfolio project research brief series*. Maker Education Initiative. https://makered.org/wp-content/uploads/2015/03/Open-Portfolio-Project-Research-Brief-Series_FULL_final-small.pdf

Charmaz, K. (2006). *Constructing grounded theory: A practical guide through qualitative analysis*. Sage.

College Board. (2017). *Advanced placement computer science principles course guide*. https://apcentral.collegeboard.org/pdf/ap-computer-science-principles-course-and-exam-description.pdf.

Conley, D. T., & Darling-Hammond, L. (2013). *Creating systems of assessment for deeper learning*. Stanford Center for Opportunity Policy in Education.

Cooper, S., Pérez, L. C., & Rainey, D. (2010). K-12 computational learning. *Communications of the ACM, 53*(11), 27–29. https://doi.org/10.1145/1839676.1839686

De Fina, A. (1992). *Portfolio assessment: Getting started*. Scholastic Professional Books.

Eris, O. (2006). Insisting on truth at the expense of conceptualization: can engineering portfolios help? *International Journal of Engineering Education*, *22*(3), 551.

Eris, O. (2006). Insisting on truth at the expense of conceptualization: Can engineering portfolios help? *International Journal of Engineering Education, 22*(3), 551. https://www.ijee.ie/articles/Vol22-3/15_ijee1806.pdf

Falkner, K., & Falkner, N. J. (2012, February). Integrating communication skills into the computer science curriculum. In L. S. King, D. R. Musicant, T. Camp, & P. Tymann (Eds.). *SIGCSE 2012. Proceedings of the 43rd ACM technical symposium on computer science education* (pp. 379–384). Raleigh, NC, USA: ACM.

Fields, D. A., Jayathirtha, G., & Kafai, Y. B. (2019). Bugs as a nexus of emergent peer collaborations: Contextual and classroom supports for solving problems in electronic textiles. In K. Lund, G. Niccolai, E. Lavoué, C. Hmelo-Silver, G. Gweon, & M. Baker (Ed.) *A wide lens: Combining embodied, enactive, extended, and embedded learning in collaborative settings*, 13th International Conference on Computer Supported Collaborative Learning, 1. Lyon, France: International Society of the Learning Sciences, 472–479.

Fields, D. A., Lui, D., & Kafai, Y. B. (2019). Teaching computational thinking with electronic textiles: Modeling iterative practices and supporting personal projects in exploring computer science. In S. C. Kong & H. Abelson (Eds.), *Computational Thinking Education* (pp. 279–294). Springer Nature.

Fields, D. A., Kafai, Y. B., Nakajima, T. M., Goode, J., & Margolis, J. (2018a). Putting making into high school computer science classrooms: Promoting equity in teaching and learning with electronic textiles in exploring computer science. *Equity, Excellence, and Education, 51*(1), 21–35. https://doi.org/10.1080/10665684.2018.1436998

Fields, D. A., Landa, J., Nakajima, T., Kafai, Y. B., Goode, J., Margolis, J., & Ottina, J. (2018b). Stitching the loop: An electronic textiles unit in exploring computer science. *Exploring Computer Science.* http://exploringcs.org/e-textiles

Fields, D. A., Searle, K. A., & Kafai, Y. B. (2016). Deconstruction kits for learning: Students' collaborative debugging of electronic textile designs. In *Proceedings of the 6th annual conference on creativity and fabrication in education* (pp. 82–85). Palo Alto, CA.

French, J. H. (2012). Evaluating a communication-intensive core course in the CS curriculum. *Journal of Computing Sciences in Colleges, 28*(2), 197–209. https://dl.acm.org/doi/abs/10.5555/2382887.2382918

Gee, J. P. (1999). *An introduction to discourse analysis: Theory and method.* Routledge.

Gilman, D. A., Andrew, R., & Rafferty, C. D. (1995). Making assessment a meaningful part of instruction. *NASSP Bulletin, 79*(573), 20–24. https://doi.org/10.1177/019263659507957304

Gitomer, D., Grosh, S., & Price, K. (1992). Portfolio culture in arts education. *Art Education, 45*(1), 7–15. https://www.tandfonline.com/doi/abs/10.1080/00043125.1992.11652907?journalCode=uare20

Goode, J., Chapman, G., & Margolis, J. (2012). Beyond curriculum: The exploring computer science program. *ACM Inroads, 3*(2), 47–53. https://doi.org/10.1145/2189835.2189851

Grover, S., Cooper, S., & Pea, R. (2014, June). Assessing computational learning in K-12. In A. Cajander, M. Daniels, T. Clear, & A. Pears (Eds). *ITICSE 14.* In *Proceedings of the 2014 conference on innovation & technology in computer science education* (pp. 57–62). Uppsala, Sweden: ACM.

Grover, S. (2017). Assessing Algorithmic and Computational Thinking in K-12: Lessons from a Middle School Classroom. In P. Rich and C.B. Hodges (Eds.), *Emerging research, practice, and policy on computational thinking* (pp. 269–288). Boston: Springer International Publishing.

Grover, S., & Pea, R. (2018). Computational thinking: A competency whose time has come. In S. Sentence, E. Barendsen, & C. Schulte (Eds.), *Computer science education: Perspectives on teaching and learning in school* (pp. 19–37). Bloomsbury Academic.

Higgs, B., & Sabin, M. (2005, October). Towards using online portfolios in computing courses. In R. Friedman (Ed.), *SIGITE 2005. Proceedings of the 6th conference on Information technology education* (pp. 323–328). Newark, NJ, USA: ACM.

Jayathirtha, G., Fields, D. A., & Kafai, Y. B. (2020). Pair debugging of electronic textiles projects: Analyzing think-aloud protocols for high school students' strategies and practices while problem solving. *Paper accepted to International Conference of the Learning Sciences*, Nashville, TN.

Jayathirtha, G., & Kafai, Y. B. (2019, February). Electronic textiles in computer science education: A synthesis of efforts to broaden participation, increase interest, and deepen learning. In

Proceedings of the 50th ACM technical symposium on computer science education (pp. 713–719). Minneapolis, MN.
K–12 Computer Science Framework. (2016). http://www.k12cs.org.
Kafai, Y. B., & Burke, Q. (2014). *Connected Code: Why Children need to learn programming*. The MIT Press.
Kafai, Y. B., Fields, D. A., Lui, D., Walker, J. T., Shaw, M., Jayathirtha, G., Nakajima, T., Goode, J., & Giang, M. T. (2019). Stitching the loop with electronic textiles: Promoting equity in high school students' competencies and perceptions of computer science. In *SIGCSE '19, proceedings of the 50th ACM technical symposium on computer science education*. ACM: New York, NY, 1176–1182).
Kafai, Y. B., Fields, D. A., & Searle, K. A. (2014). Electronic textiles as disruptive designs in schools: Supporting and challenging maker activities for learning. *Harvard Educational Review, 84*(4), 532–556. https://doi.org/10.17763/haer.84.4.46m7372370214783
Lewis, C. M. (2012, September). The importance of students' attention to program state: A case study of debugging behavior. In *Proceedings of the ninth annual international conference on International computing education research* (pp. 127–134). Auckland, New Zealand.
Litts, B. K., Kafai, Y. B., Lui, D. A., Walker, J. T., & Widman, S. A. (2017). Stitching codeable circuits: High school students' learning about circuitry and coding with electronic textiles. *Journal of Science Education and Technology, 26*(5), 494–507. https://doi.org/10.1007/s10956-017-9694-0
Lui, D., Jayathirtha, G., Fields, D. A., Shaw, M., & Kafai, Y. B. (2018). Design considerations for capturing computational thinking practices in high school students' electronic textile portfolios. In *ICLS 2018. Proceedings of the international conference of learning sciences*. London, UK: International Society Of Learning Sciences.
Lui, D., Walker, J. T., Hanna, S., Kafai, Y. B., Fields, D., & Jayathirtha, G. (2019). Communicating computational concepts and practices within high school students' portfolios of making electronic textiles. In *Interactive Learning Environments 28*(3), 284-301.
McKay, C., Keune, A., Peppler, K., Chang, S., & Regalla, L. (2015). *A networked vision for sharing and documenting - Open Portfolio Project*. Maker Education Initiative. http://makered.org/wp-content/uploads/2016/01/MakerEdOPP_full-Research-Brief-Series_final.compressed.pdf.
Moreno-Leon, J., Robles, G., & Roman-Gonzalez, M. (2015). Dr. Scratch: Automatic analysis of Scratch projects to assess and foster computational thinking. *Revista De Educaci6n a Distancia, 46*. https://doi.org/l0.6018/red/4
National Research Council. (2010) . *Exploring the intersection of science education and 21st century skills: A workshop summary*. National Academies Press.
National Research Council. (2011) . *Successful K-12 STEM Education: Identifying effective approaches in science, technology, engineering, and mathematics*. National Academies Press.
Owings, C. A., & Follo, E. (1992). *Effects of portfolio assessment on students' attitudes and goal setting abilities in mathematics*. Michiga. https://files.eric.ed.gov/fulltext/ED352394.pdf.
Parsons, D., & Haden, P. (2006, January). Parson's programming puzzles: A fun and effective learning tool for first programming courses. In *Proceedings of the 8th Australasian conference on computing education 52*, 157–163. Wollongong, Australia.
Paulson, F. L., Paulson, P. R., & Meyer, C. A. (1991). What makes a Portfolio a Portfolio. *Educational Leadership, 48*(5), 60–63. https://web.stanford.edu/dept/SUSE/projects/ireport/articles/e-portfolio/what%20makes%20a%20portfolio%20a%20portfolio.pdf
Phelps, A. J., LaPorte, M. M., & Mahood, A. (1997). Portfolio assessment in high school chemistry: One Teacher's guidelines. *Journal of Chemistry. Education, 74*(5), 528. https://doi.org/10.1021/ed074p528
Saldana, J. (2011). *Fundamentals of qualitative research*. Open University Press.
Shaw, M., Fields, D. A., & Kafai, Y. B. (2020). Overcoming challenges in e-textiles: Equitable teacher practices that support student learning and engagement in making and computer science. Computer Science Education, 30(3), 313–336
Shaw, M. S., Fields, D. A., & Kafai, Y. B. (2019). Connecting with computer science: Electronic textile portfolios as ideational identity resources for high school students. *International Journal of Multicultural Education, 21*(1), 22–41. http://dx.doi.org/10.18251/ijme.v21i1.1740

Taffe, W. (1989). Writing in the computer science curriculum. *Writing across the Curriculum, 1*(1), 17–22.

Werner, L., Denner, J., & Campe, S. (2015). Children programming games: A strategy for measuring computational learning. *ACM Transactions on Computing Education (TOCE), 14*(4), 24. https://doi.org/10.1145/2677091

Williams, J. M. (2002). The engineering Portfolio: Communication, reflection, and student learning outcomes assessment. *International Journal of Engineering Education, 18*(2), 199–207. https://www.ijee.ie/articles/Vol18-2/IJEE1268.pdf

Wing, J. M. (2006). Computational thinking. *Communications of the ACM, 49*(3), 33–35. https://doi.org/10.1145/1118178.1118215 3

Yardi, S., & Bruckman, A. (2007). What is computing? Bridging the gap between teenagers' perceptions and graduate students' experiences. In *Proceedings of the Third International Workshop on Computing education Research* (pp. 39–50).

How do students develop computational thinking? Assessing early programmers in a maze-based online game

Mariluz Guenaga, Andoni Eguíluz, Pablo Garaizar and Juanjo Gibaja

ABSTRACT

Background and Context: Despite many initiatives to develop Computational Thinking (CT), not much is known about how early programmers develop CT and how we can assess their learning.

Objective: Determine if the analysis of students' interactions with an online platform allows understanding the development of CT, how we can convert data collected into valuable insights, and the aspects that should be considered in platforms design.

Method: We developed an online platform with a fine-grained log-recording system. We analysed the data collected from 1004 students (ages 8-14) to understand the difficulties they face. We explain our platform and the tools to process and filter the interaction logs. We calculate additional indicators that provide useful information about student's behaviour.

Findings: Age and gender have shown to influence on CT learning. Generating additional indicators from basic interaction data provide valuable insights. We provide a list of recommendations for developing more effective programming learning platforms.

1. Introduction

1.1. Defining computational thinking

Computational Thinking (CT hereafter) involves a set of cognitive and problem-solving skills in which the solution can be executed by a computer (Wing, 2006). Although Wing coined the term, Papert (1980) introduced it decades before. Based on some of the well-cited works, Tang et al. (2020) grouped the definition of CT in two groups: authors that defined CT relating to programming and computing concepts, and authors who considered CT as a set of competences requiring students to develop both domain-specific knowledge and problem-solving skills. In this study we are aligned with the first definition. However, there is a lack of consensus on the definition of CT and the concepts included, thus, many challenges remain and are derived from the lack of a precise definition of CT (Denning, 2017; Wing & Stanzione, 2016).

It may also be confusing the related terms found in the literature on CT, such as computer programming and algorithmic thinking. Computer programming is the process of providing a set of instructions that communicate, as precisely as possible, a procedure to a machine. These instructions are coded using a specific programming language (Robins et al., 2003). Algorithmic thinking is a problem-solving cognitive skill that allows one to reach a solution through a series of steps. It requires analysing problems, specifying them with precision, dividing them into parts, finding the appropriate basic actions with which to solve these parts, and building a correct procedure with those actions (Futschek, 2006). Computational thinking is a broader term that includes algorithmic thinking, logic, abstraction, generalization, decomposition, and debugging, among other skills (Syslo, 2015).

Although there is no agreed-upon the definition of CT, most authors include several high-level skills and practices that are essential and applicable to many areas beyond computing (Selby & Woollard, 2013; Wu, 2018): abstraction, generalization, decomposition, algorithm thinking, and evaluation. The last one is defined as the process of ensuring that a solution is correct and that it achieves its purpose (which often requires trade-offs with speed, use of resources, or usability). However, there is less consensus on other skills and concepts included in some CT definitions (Barr & Stephenson, 2011; Grover & Pea, 2013; National Research Council, 2010), such as data/information management, systematization, logical thinking, mathematical thinking, procedural thinking, system design, debugging, parallelization, automation, modelling, simulation, recursion, efficiency, and performance limitations.

In spite of the unclear definition of CT, authors and major organizations like the World Economic Forum and the United Nations Educational, Scientific, and Cultural Organization (UNESCO) agree on the relevance of CT as part of the new literacies required for tomorrow's citizens (World Economic Forum, 2015). Therefore, there is a demand to incorporate CT into K-12 curriculum and to develop educational content and tools that makes it possible at early stages (NRC-National Research Council, 2012; Tedre & Denning, 2016).

Based on the common elements on CT definitions and the need to orient methods and tools to primary and secondary education, we created an online platform (Kodetu) that focuses on the development of CT-related skills for early programmers: abstraction, generalization, decomposition, algorithmic thinking, and evaluation. It is a closed learning environment (structured, linear and restricted in nature) that as many other learning environments have embraced the principles of game design (Israel-Fishelson & Hershkovitz, 2020).

1.2. Assessing computational thinking

The diversity we have found in the definition of CT results in the difficulty to assess it. The acquisition of computational thinking is a multi-dimensional process (Israel-Fishelson & Hershkovitz, 2020) and, as a consequence, the literature provides a great diversity of assessment types, evaluation tools and ways of operationalizing CT. Tang et al. (2020) analyzed specific types of CT assessment and four categories emerged from their work: the traditional test composed of selected- or constructed-response questions, portfolio assessment, interviews, and surveys. Based on Brennan and Resnick's framework (Brennan & Resnick, 2012), Kong (2019) differentiates the evaluation of CT concepts, practices and perspectives, using different methods in each case. In the literature review made from 2010, he identified the following methods to evaluate CT: task/project rubrics, tests designed with task-based questions, interviews, observations and reflections reports.

1.2.1 Questionnaires

Some authors have used closed-ended multiple-choice tests to assess CT. The items in these questionnaires are usually based on the visual representation of specific tools, such as Scratch (Grover et al., 2014; Lewis, 2011), Alice (Kelleher, 2006), or Kodu (Touretzky et al., 2017). Mixed closed-ended and open-ended questionnaires based on Scratch have also been used, as in the national-level tests in Israel (Zur-Bargury et al., 2013). These evaluation methods have been used both in primary and secondary education (Román-González et al., 2017), as well as in high school (Mühling et al., 2015). In general, the questions in these questionnaires include concepts such as variables, expressions, loops, and conditionals, or common misconceptions identified in the literature (Sorva, 2012). Not only visual programming, but pseudocode-based CT assessment has been used (Tew & Guzdial, 2011). More generally, Buffum et al. (2015) developed a 7-step protocol for designing, refining, and validating CT assessment instruments. Finally, a remarkable initiative is the international Bebras competition (Cartelli et al., 2010) that assesses CT through a tool-independent test (see also Gujberova & Kalas, 2013). This competition has been mentioned as an embryo for a future PISA (Programme for International Student Assessment) test in the field of CT (Hubwieser & Mühling, 2014).

1.2.2 Code analysis

The CT activities often have a computer program as output, with specific syntax in text or visual-blocks format that enables the analysis of that code to provide insights for the CT assessment. Werner et al. (2012) assessed middle school students using the Alice software and asked them to complete parts of a pre-designed code aimed at performing specific tasks. Denner et al. (2012) followed a similar approach when assessing video games created by middle school girls using Stagecast Creator. Meerbaum-Salant et al. (2011) analysed Scratch code to look for inappropriate programming habits generated by the use of this tool (e.g., inefficient use of programming structures, lack of generalization). The same authors used mixed approaches (questionnaires about code and open tasks to create Scratch programs) in a later study (Meerbaum-Salant et al., 2013). They used this approach with students in the last year of secondary school who were learning a text-programming language (Java or C#). The accuracy of code-based assessment is higher than with the use of questionnaires, but its generalization and automation are quite challenging. Also, authors particularly stressed the need to combine multiple means of assessment.

Not only the final solution -i.e., the code-, but also the programming process is crucial when developing and assessing CT. Accordingly, it is necessary to explore not only the final program submitted but the student's path towards the solution to reveal the student's misconceptions (Piech et al., 2012). Several authors have focused on the code analysis of the CT learning process. Fields et al. (2016) analysed the learning path of participants in open project solutions. Brennan and Resnick (2012) used a tool called Scrape to differentiate the types of Scratch blocks used and to convert the projects into a visual representation that quickly indicates the amount of work done and its typology. Lye and Koh (2014) conducted a meta-analysis of 27 studies (between 2009 and 2013) on CT at all educational levels. They identified many areas for improvement, such as considering the process to be more valuable than the result and more accurately logging these processes to make their analysis easier.

Measuring CT using questionnaires, interviews or observation may be arguably more authentic and lead to a deep understanding of the learning outcomes and skills developed by students. On the other hand, scoring programming projects provide valuable information but code analysis may be subjective and time-consuming (Grover, 2020). These methods do not scale to large numbers of students and, in these cases, quantitative methods based on data logs are more feasible in terms of resources.

1.2.3 Learning analytics to assess computational thinking

This is why, in the last years there has increased the use of data collected from the CT learning tools to obtain measures that could provide valuable insights on the learning process and outcomes. These studies are included in the growing area of Learning Analytics (LA). LA techniques have been used to model individual student behaviour in structured problem-solving activities, to identify patters, analyze student online learning behavior in order to improve learning outcomes, and create an understanding of how learners develop CT (Baker, 2007; Israel-Fishelson & Hershkovitz, 2020; Kelleher, 2006; Werner et al., 2013). LA has shown to be particularly suitable for large datasets.

The use of LA has the potential to provide insights into the students' CT learning process and outcomes. However, in spite of the importance of automated support for the assessment of programming activities to implement widely the teaching of computing in K-12 education, only few approaches exist (Alves et al., 2019). These authors identified 14 studies, focusing on the analysis of the code created by the students, inferring computational thinking competencies related to algorithms and programming.

Code analysis may involve static analysis that examines source code without executing the program or dynamic analysis, which focuses on the execution of the program with a set of test cases, comparing the generated output with the expected output (Alves et al., 2020). Researchers have chosen one or mixed approaches in their code analysis. Dasgupta et al. (2016) studied how remixing positively influences the learning of CT skills. They used logs of more than one million users who remixed projects. Spacco et al. (2015) logged not only the outcome but also the fine-grained edits made by students in text-based programming language (C/C++, Java, Python, or Ruby) exercises. Using 200,000 code logs from first-year college students, Piech et al. (2012) created a Markov model to describe how students find solutions to programming challenges. Vihavainen et al. (2014) used a plugin to log all the keystrokes and code snapshots during an introductory programming MOOC (Massive Open Online Course) showing how data is lost with different granularity strategies, and how this data lost varies depending on previous programming experience and the programming assignment type. Similarly, Blikstein et al. (2014) evaluated the evolution of Java and Karel the Robot source-code snapshots and found patterns with better predictive power than conventional tests.

If we focus on research with younger users, Scratch is one of the most popular tools to develop CT with young people. Block-based and game oriented, different authors have used LA techniques to assess CT with Scratch. Maloney et al. (2008) conducted a study of 536 Scratch projects, evaluating the number and type of blocks, sprites, and sounds used in them. This longitudinal study documented the increase in the use of programming concepts (loops, variables, logic, randomization) throughout the course of a year, even in the absence of instructional interventions or experienced mentors. Moreno-León & Robles (2015) developed Dr Scratch, a free tool for assessing any public Scratch project. Dr Scratch is based on two previous projects: Scrape (Wolz et al., 2011), and Hairball (Boe et al., 2013). This tool can find poor programming habits (duplicate programs, inappropriate object names, or dead code) and evaluate several features (parallelism, logical thinking, flow control, interactivity with the user, information representation, abstraction, and synchronisation). Fields et al. (2016) followed a similar approach to assess Scratch projects, but they analysed not only the final outcome but also the process through snapshots of the project every two minutes, improving the understanding of learner's developing conceptualizations of programming.

When focusing on what to assess in CT and how to convert data logged into insights about the learning, we also find different approaches from several authors. Grover (2020) focuses on measuring understanding of variables, expressions, loops, conditionals, and abstraction. She proposes the development of assessment arguments known as focal knowledge, skills, and abilities (FKSAs) targeting programming concept-related abilities for which assessment items would elicit evidence. Israel-Fishelson and Hershkovitz (2020) used Learning Analytics with participants' data logged from an online tool (e.g., progress, completion, and time on task reports) to analyse micro-persistence of students in game-based learning environments. Grover et al. (2016) combined quantitative and qualitative approaches to understand what patterns of behavior obtained from logs in a block-based programming environment provide evidence of learners' skill level and use of CT. They logged features such as the number of times a participant clicks "run" or the addition/deletion of control-flow blocks. These basic features were the basis of the statistical analysis, and ground the study of what constitutes higher-level structures and activities. They found that data-driven analyses of log data reveal evidence that the time taken to solve a level predicts skill level with ~63% accuracy, as does the time taken between runs (~55% accuracy).

Following these approaches, in this study we developed in Kodetu a Scratch-like block-based environment where participants face a set of 15 programming challenges with a limited set of possible solutions. Therefore, all studies mentiones in this article are aimed to assess CT development in terms of algorithmic thinking fundamentals, i.e, sequences, loops, one-branch conditionals, two-branch-conditionals and code size limitations.

Pelánek and Effenberger (2020) propose three levels of granularity to log users' activity. In this study we opted for the most detailed one, which involves registering each change on the workspace. More granularity means larger memory requirements, but more potential for data analysis. Kodetu integrated a fine-grained logging system that records every interaction of participants with the platform, along with the timestamp, userID, code-blocks in the solution and other features described later. As shown in previous works, user's performance and progress through the CT tasks can be measured with the data obtained automatically from the tool, as well as with the new indicators obtained aggregating data.

1.3. Objective and research questions

The analysis of automatically collected data from an online platform for the development of CT provides a better understanding of the learning process and outcomes, as well as the users' performance and other indicators for analysis. Logging the activity of thousands of users in a single session interacting with an online platform yields information that we could hardly obtain by other means. However, if these data is not properly processed using computers and the appropriate algorithms it could limit the depth of the knowledge or the interpretation of the results obtained. Considering this, the objective of this study is: to analyse the log-data provided by a maze-based online game to generate indicators that allow us to have a better knowledge of the outcomes and process of the acquisition of CT in early programmers.

To understand how students develop CT, we designed, developed, and tested the maze-based online game called Kodetu. Even it has been available online for the public since 2018, we conducted a series of controlled workshops with primary and secondary school students to answer the following research questions:

RQ1. Does the analysis of a student's interactions in Kodetu allow us to better understand the progress of CT performance?

RQ2. How do personal features (gender, age, attitude towards technology, and prior programming knowledge) influence early CT experiences?

RQ3. How the indicators that can be defined in a log-based system influence the performance of students in this particular type of system?

These research questions will be answered from the partial view offered by a log-based analysis. We admit that trying to understand the development of CT only from these data is limited as social and psychological aspects of learning are not gathered. However, as we will explain in the following section, this is the only evidence we could capture in a procedure that involved more than 1000 participants in different workshops.

2. Methods

To answer our research questions, we developed two complementary tools: a public online platform, Kodetu, and a desktop data analysis tool, KodetuWin.

2.1. Kodetu

Kodetu is a challenge-based online platform designed to convey some of the most basic programming concepts: basic sentences, sequential structure, loops, and conditionals.

Kodetu is based on Blockly, a software library developed by Google to create block-based visual programming environments, particularly on Maze, which implements a series of maze levels which challenge users to find the exit using visual blocks of code. We modified this software in three ways: (1) We increased the number of levels to 15. More advanced levels involve a higher challenge, (2) We added a detailed log system that collects all the students' actions while using the platform, and (3) We developed an online platform that allows user identification, group management, and different setups for each group.

At the beginning of each Kodetu session, all students are informed about the features and objectives of the research that we are conducting and are asked for their voluntary participation in the study. Considering that we are dealing with underage students, we established a hidden identification system that assigns a unique random alphanumeric code to each participant. To ensure privacy and anonymity, we request no personal information and do not store IP addresses.

The system asks for demographic data (age, gender, and educational level), the name of the school, and two questions related with the topic: whether the student has previous knowledge of programming (yes/no), and the student's level of interest in technology (a numerical value from 1 -low- to 10 -high-).

Figure 1. An example of a Kodetu maze.

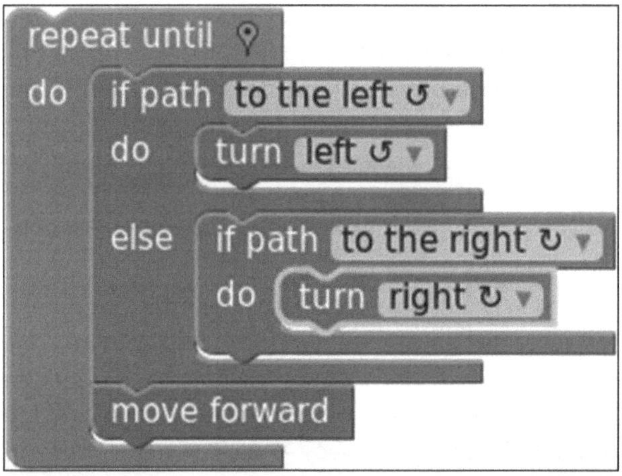

Figure 2. Example of code with Kodetu blocks.

After these data are entered, Kodetu shows a maze in the form of an orthogonal path of a schematic space station (in a virtual 8 × 8 grid, Figure 1). An astronaut is located at one point in the maze and a single goal mark is placed at another point in the maze. To reach the goal, the astronaut must walk on the surface of that space station while avoiding falling into space (i.e., going off track).

To move the astronaut, students are allowed to use a reduced set of code blocks (Figure 2) representing basic movements: move forward (fw), turn left (rl-rotate left), turn right (rr-rotate right). Later in the game, other blocks will be available: repeat until, if path {forward, to the left, to the right}.

The Kodetu interface has a main area divided into three parts: the left displays the maze(s) to be solved and the "Run Program" button; the center displays the blocks available for programming; and the right displays the workspace (Figure 3).

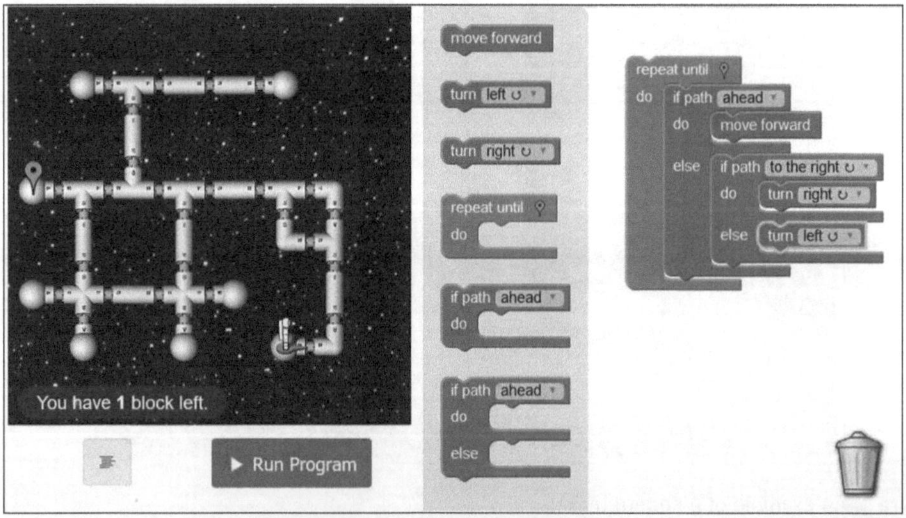

Figure 3. Kodetu interface.

Students need to select and move blocks from the central area to the workspace in order to build a program which takes the astronaut from the start to the exit. When they consider that they have found the solution using the programming blocks, or any time they want to test it, they can press the "Run Program" button. This makes the astronaut follow the movements defined by the blocks in the workspace. If the astronaut falls from the space ship or stops before reaching the goal, an error message is displayed. Then, the students can change their code and retry. If the astronaut reaches the goal, a new, more complicated challenge is presented.

The 15 Kodetu levels are divided into four types: A) Sequences (levels 1 to 6), B) Loops (levels 7 to 10), C) Loops with simple alternatives (levels 11 to 12), and D) Loops with double alternatives (levels 13 to 15). Level 1 presents an initial challenge as a tutorial and its solution involves only two forward blocks (fw). Level 2 includes a right turn (rr), so students learn how the rotation blocks work and how this rotation is relative to the direction in which the astronaut is looking (Figure 4). Starting at level 3, students must solve a dual maze with a single code. Solution: [rr fw fw]. Levels 3 to 6 present mazes with symmetric paths and cases in which one of the mazes requires more blocks than the other. From levels 7 to 10, loops (rep) are introduced. In these levels, students must find the paths to be repeated and the similarities between both mazes. Levels 11 and 12 introduce simple alternative blocks to determine whether there is a path to the left (if-l), to the right (if-r), or forward (if-fw). Students will combine the previous loop block with this simple alternative block in these levels. Levels 13 to 15 introduce a new type of block, the double alternative (if [] else []). Only a few students can solve the complex mazes presented in the last level (Figure 4).

Based on previous research on the CT assessment, log-based and other methods (Brennan & Resnick, 2012; Grover, 2020; Grover et al., 2016; Israel-Fishelson &

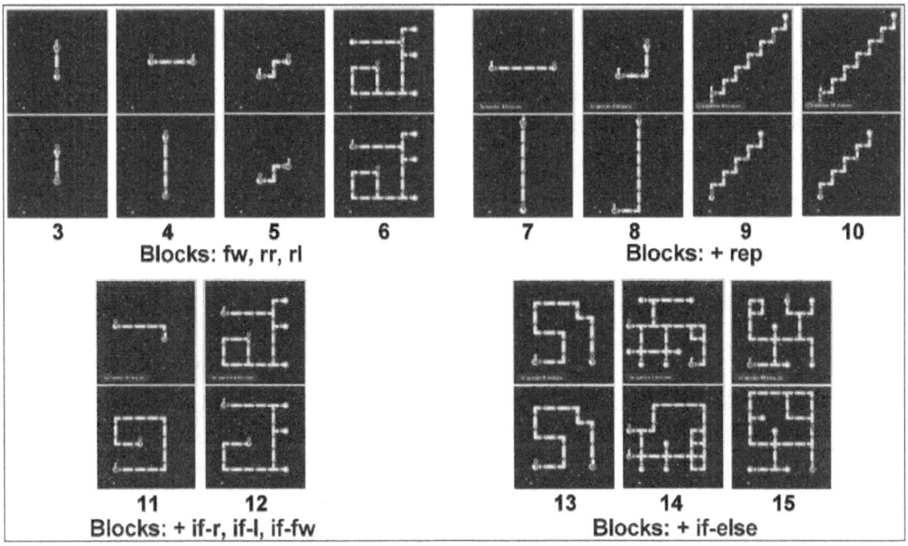

Figure 4. Kodetu mazes 3 to 15 grouped according to the four possibilities depending on the available blocks.

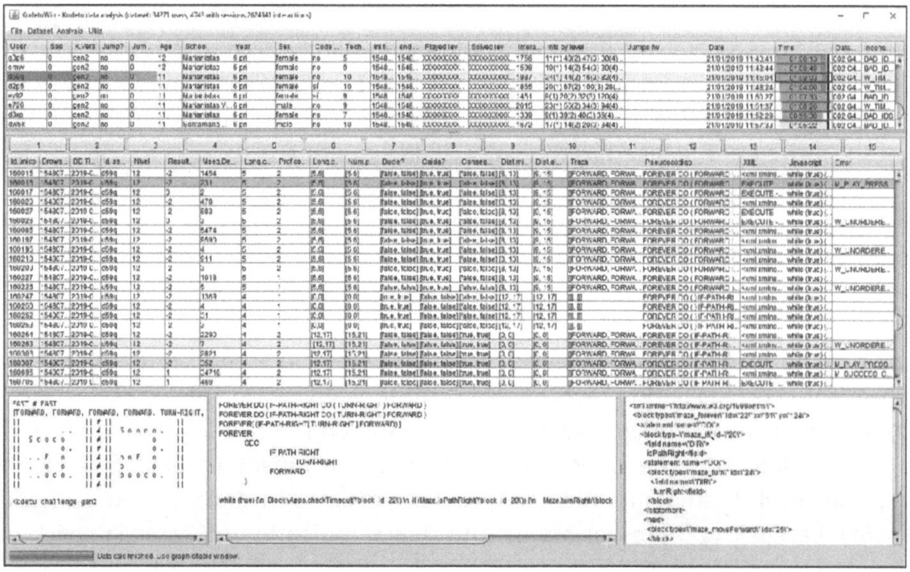

Figure 5. KodetuWin interface.

Hershkovitz, 2020; Pelánek & Effenberger, 2020) we defined which data automatically collected through the CTTOLL provide valuable information for analyzing the development of CT and student's performance. Accordingly, Kodetu stores: the timestamp, the userID, the level played, the code in the workspace, and a number representing the outcome solving the challenge (0: no attempt, the execution button has not yet been pressed, 1: success, the astronaut has reached the goal, −1: failure, the code ends without reaching the goal, −2: fall, the astronaut falls into space, or 2: infinite loop, the path is repeated indefinitely without solving the maze).

2.2. KodetuWin

KodetuWin is a specific purpose tool for processing Kodetu logs (Figure 5). We developed this Java application to browse and analyse each student's session (upper panel), the Kodetu levels, and each student's interactions -block changes and executions- in each level (central panel).

KodetuWin not only simplifies Kodetu's logs parsing but also provides interesting extra features. It includes an execution simulator that allows researchers to simulate the execution of any student's code on the corresponding maze. Moreover, KodetuWin shows researchers the path the astronaut would follow for each code in the workspace even if the student did not execute that code. This is particularly interesting in cases in which the students did not realise that they had successfully completed the challenge.

Not only raw data logged from participants' interactions are useful for the assessment of CT. We can obtain higher level measures combining and aggregating these data. We based our indicators in previous works that defined the measures that can be extracted from logs and provide information about, for example, the difficulty of a puzzle game (i.e.

failure rate, the median time to solve the puzzle, and the median number of actions like edits, executions, or submits) and the complexity of the challenge (i.e. the number of programming concepts involved, the number of flow-of-control structures), which relate with the performance of users in a block-based programming environment (Pelánek & Effenberger, 2020).

Accordingly, KodetuWin calculates metadata for each user interaction:

Time elapsed	Time elapsed since the previous interaction
Code difference	Number of blocks removed, modified or added since the previous interaction
Code length	Number of code blocks in this interaction's workspace.
Code depth	The highest level of nesting of the control structures.
Outcome	No attempt, success (astronaut arrives to goal), failure (code stops before goal), fall (astronaut falls from the spaceship), or infinite loop (goal not reached but code never stops).
Path length	Length of the path that the astronaut follows when attempting to reach the exit.
Number of movements	Number of movement blocks executed in the maze (turns or moves forward).
Number of dead code blocks	Blocks that are not executed because the execution ends or loops before reaching them
Cost of execution	Number of blocks that are executed, including conditions that are checked in alternative structures.
Visual distance to the goal	Manhattan distance (the distance between two points measured along axes at right angles) in tiles from the goal with respect to the path taken, not depending on which tiles are in the middle.
Distance from the path to the goal	Number of valid steps that would remain from the path taken to reach the goal.
Trace of the path taken	List of tiles the astronaut walked on.
Trace of the code executed	List of blocks executed and conditions tested

Based on these data, cumulative indicators are also generated by level and student:
Interactions per levelNumber of interactions per level.Time per levelTime spent in the level, from start to solve/finish time. Attempts per levelNumber of times the student presses the "Run Program" button while trying to solve that level.Code difference per levelNumber of blocks deleted, modified, and added.Level solution code(if any) Final code with all associated values (length, depth, path length, number of executed blocks, dead code blocks, execution cost, distance to the goal, path trace, code trace).

KodetuWin helped us enrich our data set to better assess insightful aspects of students' interactions from a CT perspective, such as unattempted solutions (code that would have solved the challenge but that the student did not try to execute) or proximity to the solution of the student's code.

Due to the overwhelming number of variables that KodetuWin provides, we decided to develop a preliminary real–time visualization system, which allows us to graphically see the aggregated values and main statistics (mean, median, standard deviation) for the variables on any filtered data set (Figure 6).

To avoid spurious data, we considered only the first session of each student in this research, since in the following sessions (second and beyond) the learning acquired in the first one interferes with the measurement of the user's performance. The same student can have different Kodetu sessions recorded since they may restart the browser or reload the page, and it provokes a new session to start in the platform. A Kodetu session comprises all the consecutive interactions that occur in the platform, starting at level 1 and it ends when a level is skipped or after 30 minutes of inactivity. To split sessions, KodetuWin uses the timestamp of the interactions.

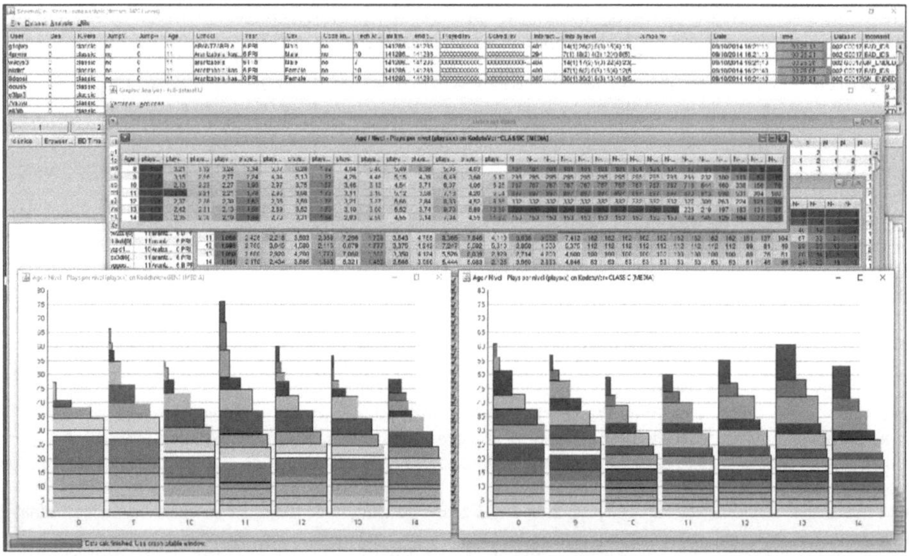

Figure 6. Example of aggregate data tables and graphics in KodetuWin.

2.3. Participants

In total, 1,004 students from 8 (3rd grade of primary school) to 14 years old (2nd grade of secondary school) generated 732,478 Kodetu interactions, logged between April 2018 and January 2019 in controlled workshops of 20–25 students.

2.4. Procedure

The Kodetu platform has been publicly active and open since March 2018. In many cases, uncontrolled users play the game from anywhere in the world. For our study, we organized controlled workshops with primary and secondary schools that were part of a larger visit to the Faculty, our labs and other workshops. Kodetu workshops were carried out during school hours with one computer per person. We introduced students to the experiment and asked them to solve the Kodetu challenges without help. The usual duration of these workshops was 50 minutes, although for various reasons it was not possible to set the same time for every student.

Once we had conducted all the workshops, we used KodetuWin to filter and verify our interactions data set in the following ways:

(1) Success outcomes. All the outcomes marked as success were marked as success were automatically verified to ensure that the logged code effectively solved the corresponding maze(s). If not, these sessions were discarded.
(2) Level jumps. We checked the timing sequence of levels for each student. Some students jumped to previous levels using their Internet browser's back button, and communication errors caused loss of some interactions. Sessions with any of these errors were discarded.

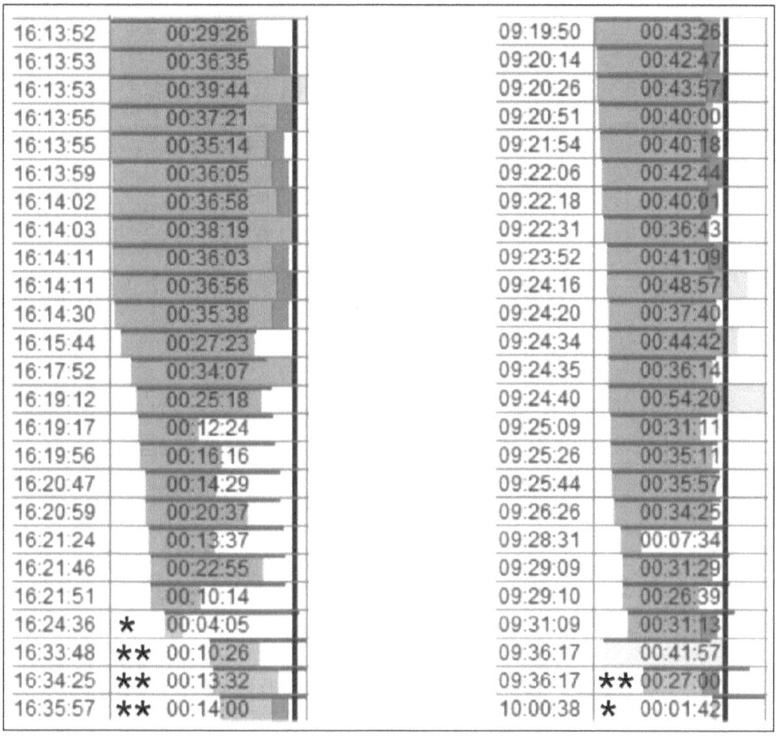

Figure 7. Group session timetables with group-end lines, extended sessions, and marks.

(3) Session consistency. We filtered and discarded sessions with long time breaks (over 30 minutes between interactions), too-short sessions (fewer than 10 interactions or only levels 1 and 2 completed), or not enough data logged.
(4) Controlled workshops. Most student groups taking part in our workshops were identified using their group codes. We also designed an algorithm to identify the group of students who accessed Kodetu concurrently, according to the start and end times of the session and the name of the school. These are identified as a possible group by KodetuWin. Then, we checked students' ages and schools to verify that it was, indeed, a group. At this point, we split the data set in two – controlled groups and uncontrolled groups – and discarded all uncontrolled groups' sessions.

During the workshops, some students had technical problems or personal incidents (e.g., they arrived late, could not find the classroom, went to the bathroom, or had to leave early); these students did not have the same session time as their classmates. Considering this, we decided to identify these cases by calculating the session time of each student within the group, as these circumstances can significantly affect performance.

First, we calculated the session time for each student within a group. Table 1 represents this information, where the width of the Time column represents the total time spent by the group in the workshop (from the first start to the last end) and the grey shade represents the time of the student.

Table 1. Shaded, student time within the session.

Date	Time
13/04/2018 16:13:48	00:36:08
13/04/2018 16:13:48	00:30:46
13/04/2018 16:13:48	00:19:39
13/04/2018 16:13:49	00:37:01

Second, we sorted the students by their start times and found a few cases in which timing discrepancies could influence our research: 1) time translated sessions (caused mostly by unsynchronized computer clocks); 2) extended sessions (students' sessions lasting significantly longer than their peers'; see at Figure 7 the end of the group marked with a vertical dark line, extended sessions in light grey after the line); 3) short sessions (see Figure 7, one star); and 4) late sessions (those starting over 15 minutes after the beginning of the workshop; see Figure 7, two stars). All these sessions were discarded from our data set.

Finally, we calculated aggregated measures for each session, dividing levels into programming concepts broadly used in the CT assessment literature, these are: elementary commands, loops, conditions, and integration skills like the use of nested structures (Kong, 2019; Pelánek & Effenberger, 2020). We grouped the 15 levels of Kodetu considering these CT concepts included and the blocks available:

- Type A, levels 1–6: Levels that are solved using only sequences of blocks (move forward, turn right, and turn left: fw, rr, rl).
- Type B, levels 7–10: Levels that require loops (rep) and sequences to be solved.
- Type C, levels 11–12: Levels that require simple alternatives (if), loops, and sequences.
- Type D, levels 13–15: Levels that require single and double alternatives (if-else), loops, and sequences.

We discarded level 1 and level 7 from the analysis because they are considered training levels (level 1: how to use Kodetu; level 7: how to embed blocks into other blocks). Thus, for each student, we calculated the following aggregated values for each Kodetu type:

- **Proportion of levels solved for each type of level**. As the number of levels in each type is not homogeneous, we normalized it in the [0, 1] interval. When all the levels of a type are solved, the value is one.
- **Average number of plays per solved level**. Considering all the levels solved within each type, we calculated the average number of plays (clicks on the "Run Program" button) for each student. Obviously, this value is at least one, as it is necessary to press the "Run Program" button at least once to solve the level. In each case, we

calculated this value as the number of "Run Program" clicks on the levels of the type, divided by the number of levels solved of that type.

- **Average of total time per solved level**. Similarly, we calculate the total time each student spent in all the levels solved of a type, divided by the number of levels solved. We also calculated the inverse value: the number of solved levels per minute, which is sometimes easier to understand.
- **Average plays per minute in solved levels**. We wondered about the relevance of measuring how often the students tried to execute their code. Therefore, we calculated the ratio between the number of plays in all solved levels of each type and the time spent in those levels (in minutes).
- **Average extra code changes**. We calculated the minimum number of changes in the code needed to solve each level (i.e,, the difference in length between the initial code and the code with the final solution assuming no deletion or modification is made). Thus, we calculated the difference between the changes made by the student and the minimum changes needed. Therefore, this value is zero when the student made the minimum number of code changes necessary to solve the level, and positive otherwise. We also calculated the average of all the extra changes made in the solved levels of each type and the number of levels solved.
- **Ratio of code length and the smallest correct code length**. We calculated the ratio between the sum of the code length of the solution of the levels of each type and the sum of the lengths of the shortest codes that solve those levels. This value is 1 in the best case and higher otherwise.

3. Results

Our approach to the study of the development of CT is based on the evaluation of each participant's performance at each level individually. Since we take into account both the number of attempts and the time spent to solve a level, we consider that this analysis penalize trial and error strategies and also strategies that require a long thinking time before making their first attempt. Although both strategies are acceptable and denote an effort towards the task, they also indicate a lack of skill at that level. In this way, we estimate the development of CT according to the evolution of the participants' performance in each of the levels individually.

3.1. Game time analysis

Table 2 shows the relationship between the duration of the game session (in 5-minute intervals) and the level finally passed. The average level achieved progresses linearly (Figure 8) with respect to time (K-W: chi-square = 349.57, df = 13, $p < 0.001$).

Although most of our students are in the 21- to 45-minute range, these data show that playing Kodetu may require more time than that offered in our workshops. We cannot determine the ideal duration of the session, as we have not carried out sessions in which we could observe better results with higher available time.

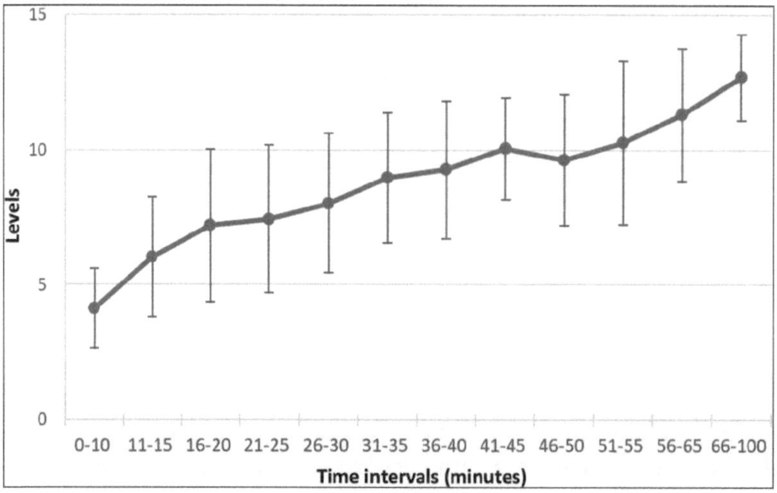

Figure 8. Evolution of the average level passed with respect to the time played (intervals in minutes).

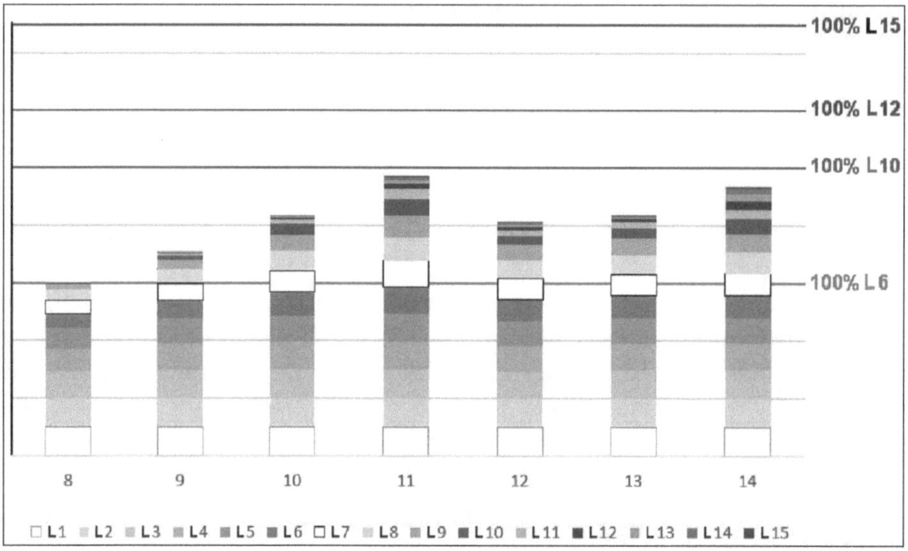

Figure 9. Cumulative percentage of students (Y-axis) who passed the levels (X–axis) per age. Each block represents a level, increasing from the bottom to the top. The white blocks are the training levels (1 and 7).

3.2. Analysis by age

The increasing difficulty of the levels and the time limitation cause a downward progression in the levels passed from 1 to 15. If we split the sample by age, we expect a positive relationship between age and performance. However, this growth is not linear. There is a particularly good performance at age 11 (average level exceeded 9.7). Figure 9 shows the accumulated levels passed by age, where we can see this growing progression.

Table 2. Heat map of the level achieved as a function of playing time.

	N	Level succeded (% on initial users) 1	2	3	4	5	6	7	8	9	10	11	12	13	14	15
0-10 mins	60	100%	100%	85%	62%	45%	12%	8%	2%	0%	0%	0%	0%	0%	0%	0%
11-15 mins	57	100%	100%	96%	88%	81%	47%	35%	26%	19%	9%	2%	0%	0%	0%	0%
16-20 mins	70	100%	100%	97%	89%	86%	63%	56%	44%	40%	24%	16%	1%	1%	1%	1%
21-25 mins	102	100%	100%	95%	91%	84%	70%	65%	58%	41%	23%	10%	3%	2%	2%	1%
26-30 mins	210	100%	100%	99%	97%	93%	79%	74%	60%	45%	28%	12%	8%	4%	3%	2%
31-35 mins	190	100%	100%	99%	99%	98%	89%	84%	79%	66%	38%	17%	10%	8%	6%	4%
36-40 mins	127	100%	100%	98%	98%	97%	90%	89%	82%	67%	48%	24%	16%	11%	7%	2%
41-45 mins	96	100%	100%	100%	100%	99%	98%	97%	92%	88%	65%	35%	17%	9%	5%	2%
46-50 mins	38	100%	100%	100%	95%	95%	89%	89%	87%	79%	66%	37%	18%	8%	0%	0%
51-55 mins	15	100%	100%	93%	93%	93%	93%	87%	87%	87%	80%	53%	33%	13%	7%	7%
56-65 mins	25	100%	100%	96%	96%	96%	96%	96%	96%	96%	84%	76%	60%	28%	8%	4%
66-100 mins	14	100%	100%	100%	100%	100%	100%	100%	100%	100%	100%	86%	71%	64%	50%	0%

The dropout rates (participants stop playing) show that 86.9% of 6th primary school students finish the sequential levels, compared to 82.3% in the 1st year of secondary school and only 69.0% in the 2nd year of secondary school.

To further explore this difference, we look at effective play time (the real time that each student spends solving levels), through which we can see a much higher engagement of 11-year-old students who use many more effective minutes of play than their older peers (average 30.3 minutes versus 23.1 or less). To discard that 11-year-old students indeed perform better than older students, we have calculated a specific indicator that is not affected by motivation or dropout: the number of levels passed per minute. In Figure 10 we can see that the progression solving Kodetu challenges is linear and correlated with age (11-year-old students perform better not because they are better but because they try longer).

We analysed the time used per passed level at each age (Figure 11). It shows the average time that students of each age spent in each of the 15 levels (in minutes) when they passed it. In columns, we show the relative survival at each age and level, and observe how it decreases as the game progresses. We can observe that the youngest students do not even get to play the higher levels (no 8-year-old student passed level 12). We can also see the natural decreasing trend of time spent by age in each level. We can see some particularities in the progress of levels that would help us to carry out a pedagogical redesign of the tool, such as the greater relative time spent in level 6 (solution built only with a sequence of movements) or the relative difficulty of level 11 (introduction of alternative structures). Also, we observe the behaviour of the 11-year–old group in this table: they passed levels 13 and 14 thanks to their extra effort and they are the students who spent more time in levels of type D (double alternatives in levels 13–15). These differences are significant in time at levels 1 to 12 (K-W: p between 0.02134 and < 0.001) and non-significant at levels 13 to 15 (K-W: $p = 0.09634$).

We can see how the more demanding levels (higher in the figure) require less average time at older ages. In this sense, we could identify 14 years as the most appropriate age in

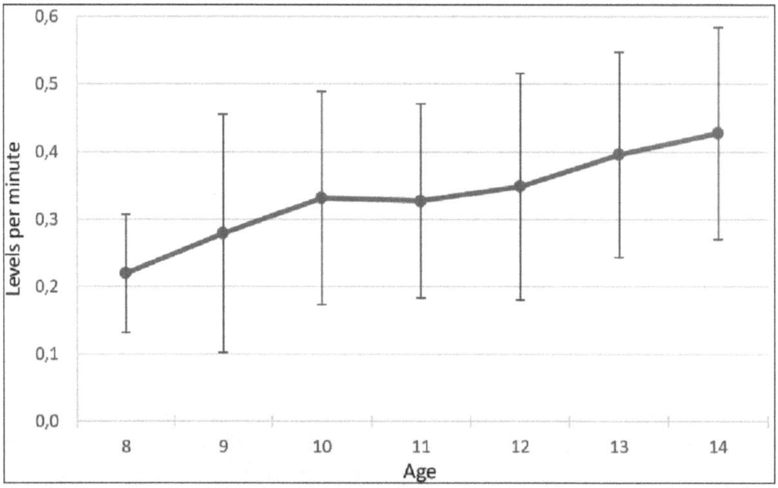

Figure 10. Levels passed per minute, per age.

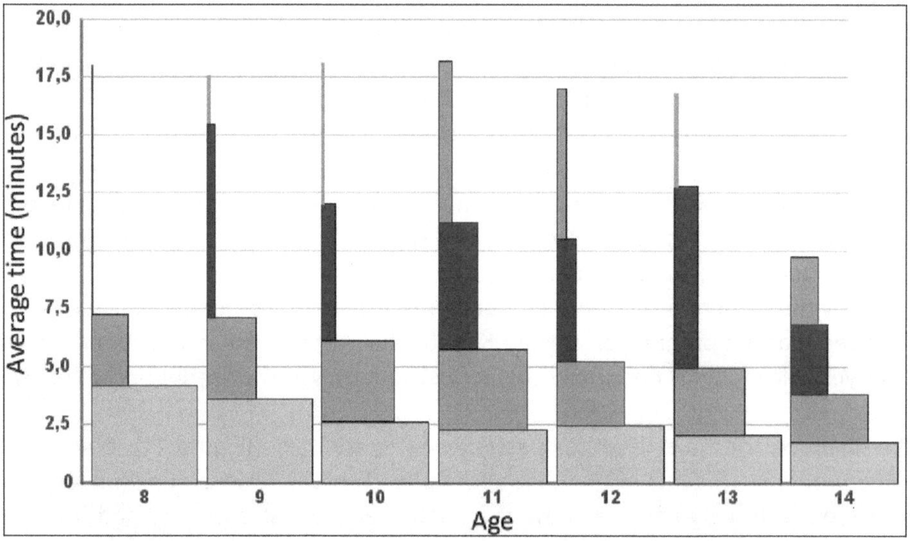

Figure 11. Average time in each level passed by age. Boxes stacked vertically by Kodetu type, from A (bottom) to D (top). Box width: percentage of students remaining at each level and age.

this range for facing Kodetu challenges, although the whole range from 11 to 14 years can face these challenges.

3.3. Analysis by gender

Our sample consists of 519 girls (51.7%) and 484 boys (48.3%). The gender distribution is not homogeneous by age; it ranges from 44.3% of girls in the 8-year-old group to 66.7% of girls in the 12-year-old group.

In terms of the most relevant indicators, boys perform slightly better than do girls at the intermediate levels; the difference is less, and even reverses, at level 15. Girls perform better at level 15 despite being younger (the average age of the 9 girls who passed level 15 is 11.56 years, while the 13 boys who passed level 15 have an average age of 12.38). The average level passed is 8.5 for boys and 8.2 for girls (M-W-W: W = 2,456,578, $p < 0.001$). Boys use less time than girls (5.2 minutes difference), except at level 15, where girls are faster (2.6 minutes faster) (M-W-W: $p < 0.05$ at levels 1 to 12 and 15). Boys make fewer attempts to pass levels (3.1 fewer than girls) but at level 15 this reverses (girls make 4.8 fewer attempts than boys) (M-W-W: $p < 0.05$ up to level 9). Boys also make fewer code changes (424.6 versus 433.5) except at level 15 (47 changes plus boys) (M-W-W: $p < 0.05$ up to level 7).

3.4. Analysis by programming knowledge

We analyse the same indicators, relating them to the previous knowledge of programming declared by the students in the questionnaire before the experiment, "do you have previous knowledge of programming (yes/no)". In total, 79.6% indicate no previous knowledge, while the remaining 20.4% do. As we might expect, the difference is very high: 92.9% of the younger students (8-year-old) have no previous knowledge, while 80% of the oldest have some programming knowledge.

As expected, we can see that students who declare prior knowledge of programming perform better. The average level they reach is 8.9 -with knowledge- compared to 8.2 -without- (M-W-W: W = 1,454,884, $p < 0.001$). This difference increases at the levels of structured blocks (reaching a difference of 13.22% at level 11). It makes sense if we consider that previous experience will often provide insight into how to use repetitive and alternative structures.

3.5. Analysis by attitude towards technology

To analyse the students' results in relation to their attitude towards technology we focused on the question they answered before the experiment ("How much do you like technology?"). The answers were rated between 1-minimum to 10–maximum. We calculated the Spearman coefficient and saw that there are no consistent correlations between student attitude and the progress, time spent, executions, or changes in the code. This result could be caused by low variation in recorded values on this question.

Only 26 students (2.6%) used values 1–2 to assess their attitude towards technology; 67 (6.7%) used values 3–5; 293 (29.2%) used values 6–8; and 618 (61.6%) used values 9–10. In other words, over 90% of the students declared a 6–10 affinity, and almost half (47.8%) answered the maximum value.

Considering that the subject we are developing is CT and that the tool is technological in nature, we could expect a better performance from the more technology-oriented students. However, results do not support this assumption. Participants who rated their attitude as 6–10 all performed similarly, and only the 10% of participants who rated attitude below 6 passed fewer levels.

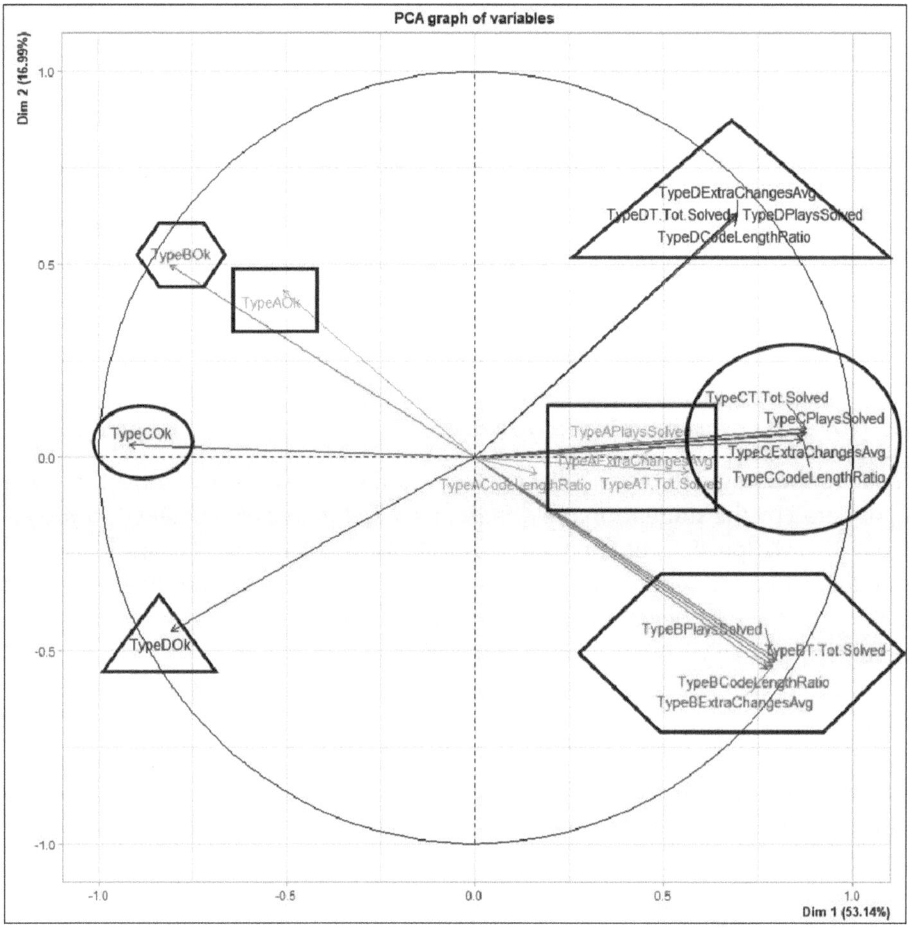

Figure 12. PCA analysis of all the primary and secondary variables, marked with geometrical forms by type.

3.6. Performance analysis: PCA

Given the large number of data available and in order to calculate an overall performance, we studied different ways to structure this data set. Not wanting to reduce the number of records, we chose to reduce the number of variables, finding variables that look alike based on their correlation. For this aim, we apply the unsupervised learning technique PCA (Principal Component Analysis).

In this way, we analysed the overall performance of each participant, applying the PCA analysis to the variables used (see Figure 12). It clearly shows the inverse correlations between each level-pass variable (TypeAOk, TypeBOk ...) and the secondary variables (TypeAExtraChangesAvg, TypeAPlaysSolved, TypeAT.Tot.Solved, TypeACodeLengthRatio, etc.). In the figure, we have marked each variable by its Kodetu type; we can see that type A (rectangles) is the one in which the variables have a lower correlation. The others are almost straight lines in only one direction.

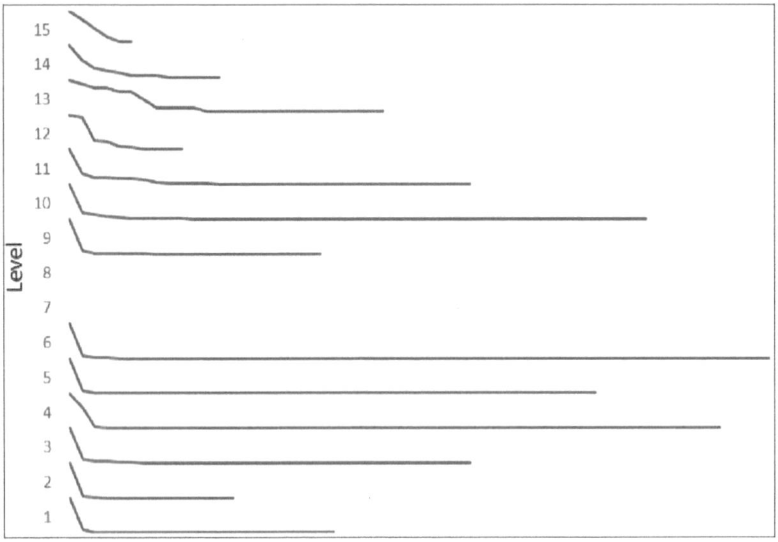

Figure 13. Long queues of code variants (horizontal) for each level (vertical). Each line represents the percentage of students who repeat each unique solution, from left (most common codes) to right (less common codes). Longer lines indicate more different codes.

In the analysis, we have seen a fundamental variable – the passed level index – which increases in the same direction of performance (the higher the level passed, the higher the performance). On the other hand, four variables that have an inverse relationship (the lower, the better) are time spent, number of executions, number of extra changes, and code length ratio. Actually, these secondary variables allow for a deeper analysis of the way students program, although the trend in the group has an inverse relationship to performance. Based on the PCA analysis and the weight provided by each variable, we defined a formula that integrates every value with the coefficient calculated from the PCA. Considering the variables and studying their relative importance in order to make this classification, we understand that a significant weight must be added to the primary variable, index of levels passed, with which we add a weight of 16 for it, maintaining a weight of 1 for the secondary variables. To incorporate these variables, as they are inversely proportional to performance, we subtract them from the maximum value for each group. The scoring formula is:

$$16. \sum_{t\in[A,B,C,D]} k_t \cdot ok_t + \sum_{t\in[A,B,C,D]} ktime_t \cdot (Ltime_t - time_t) + \sum_{t\in[A,B,C,D]} kplays_t \cdot (Lplays_t - plays_t)$$
$$+ \sum_{t\in[A,B,C,D]} kchang_t \cdot (Lchang_t - chang_t) + \sum_{t\in[A,B,C,D]} kratio_t \cdot (Lratio_t - ratio_t)$$

(where Ki is the coefficient and Li is the value limit)

3.7. Analysis of code

When developing CT, it is essential to consider the fact that there can be several correct solutions to the same problem. We analysed the code that students built to solve the different levels, and we found a particular characteristic: A large percentage of students chose a few common solutions, and the other solutions extended to many different variations (Figure 13). Despite the limited context that Kodetu provides for code expressiveness (in some levels, the expressiveness is impossible), we found relevant code variants in several levels.

In levels 2 to 6, despite having only sequential code blocks, we found many variants. Between 56.7% and 93.8% of the solutions (depending on the level) were the expected ones. The unexpected solutions included dead code and lateral variants (i.e., [rr rr rr] instead of [rl]). In level 4, in addition to the optimal six-step codes [rl rl fw fw fw] and [rr rr fw fw fw], there were up to 50 different solutions (provided by the 96 students who did not provide the two most frequent solutions). At level 6, where several paths of different efficiency can solve the challenge, we found up to 57 different solutions (the least efficient one uses up to 33 more blocks of code than the minimum 15 blocks required).

The greatest expressiveness and richness of code is from level 11 on. At level 11, 50% of the students found the best solution and there were 31 other solutions. At level 12, despite its code limitation to 5 blocks with a minimum code length of 4, there were 10 solutions and the most frequent one (rep [if-fw [fw] if-r [rr]]) was not the shortest. Equivalences of code in these levels are not so obvious and do not necessarily indicate incorrect code, but they are interesting combinations of alternatives and sequences resulting in identical paths, or variants of turns that combine in different ways.

Level 13, where the code length is between 6 and 10 blocks, is the one with the highest diversity of solutions in Kodetu. The most common solution was used by only 13.4% of the students, and we found 25 other solutions. Level 14 is similar, but with a smaller range of variety because the code is limited to 6 or 7 blocks. Only 31.1% of students chose the most common solution.

When we designed levels 13 and 14, we considered that some solutions to level 13 would also solve level 14. We analysed the algorithms that solve both levels, finding that 50% of the students solve level 14 with the same code used in level 13. It is important to emphasize that the code is not stored in Kodetu and must be rewritten from scratch at each level. Also, 12.9% of students whose solution to level 13 does not solve level 14 try to repeat the code without success.

Repeating code between levels also occurs at level 15. In total, 17.8% of the students wrote code at level 14 that was so general, it could also be used to solve level 15.

This wide variety of possibilities, in a context in which a different code is not necessarily worse, allows us to classify students according to the originality of their code, considering the number of students who found the same solution as a percentile. We also use this data in our analysis. We wonder whether the students with the most original code are always the same. There seems to be a relationship between levels 1 and 7, but it changes between levels 11 and 15. Few students have original codes in both series, which shows that the approach used in pure sequential programming is significantly different from that allowed by algorithms with control structures.

Another interesting level to study is level 10, which was intentionally left without limits in the experiment. It should be noted that level 10 is presented after three levels working with repetitive structures– two of them with the maximum restriction (levels 7 and 8) and another one with limitation but expressiveness (level 9: 8 blocks maximum, there are solutions with 5 to 8 blocks). The block limit information is not explicit at the beginning of the level (there is a small text that provides information about the number of blocks available). We found three relevant results:

1) Although levels 9 and 10 are very similar (10 needs one more block than 9, an initial turn), there are many differences in the solutions. At level 9, there are 21 different codes provided by 550 students (3.8%), while at level 10, there are 47 solutions programmed by 370 students (12.7%). The average solution has 5.2 blocks at level 9 and 8.9 blocks at level 10.

2) In total 8.9% of students re-use sequences and ignore repetitive structures to solve a level that requires at least 18 blocks without loops.

3) 13.8% of students use repetitive structures with undesirable solutions: repetitive blocks that include twice as many blocks as needed, or nested repetitive ones that make no sense in Kodetu.

4. Discussion

Effective CT learning requires students to know their level of performance on a task and how it relates to good levels (Sadler, 1989). Performance assessment of CT seems to have some advantages compared to traditional methods due to the capacity to assess higher-order thinking (Torrance, 1995). However, it results in incomplete insights and understanding of CT acquisition if it comes only from log-data analysis. At the same time, it eases teachers' workload, eliminates bias and ensures consistency and accuracy of assessment results (Alves et al., 2019).

A system such as Kodetu, which logs the interaction of users in each action, enables the collection of a great amount of data and the generation of more complex indicators the analysis of which leads to better understanding of the CT development. Using Learning Analytics techniques authors have analysed the design of the challenges and their scaffolding (Baalsrud Hauge et al., 2015); the users' behaviour, the result of their effort and, more relevant, the process to achieve that result (Ruipérez-Valiente et al., 2015); also, we can see the different learning strategies, and how learners deal with error or dropouts (Cambruzzi et al., 2015). In these cases, and also when applied to CT assessment, LA techniques in a large sample of users lack a holistic view and understanding of CT acquisition and do not answer questions related to the reasons underlying student's performance and results.

With Kodetu, researchers can design CT challenges and control the elements that change between one level and the next one (e.g., introducing a new instruction, increasing the size of the maze or limiting the number of blocks in the final programme). Analyzing the change in indicators between the successive levels, and the overall progression, we can determine how the variations introduced in the design of the challenges affect the indicators measured.

In the previous section, we presented the indicators generated from the users' interaction with the Kodetu, and their analysis allows us to answer the research questions in this study.

RQ1. Does the analysis of a student's interactions in Kodetu allow us to better understand the progress of CT performance?

The analysis of students' interactions with a system like Kodetu provides insights for an exploratory analysis of the development of CT. The fine-grained and systematic logging of the game sessions allows us to analyse the codes that students try and how students progress as they face errors and different challenges, as well as to observe how the different groups perform. Kodetu allows the exploratory analysis of CT development (de Souza et al., 2019), in particular, that related to the performance of the students, the analysis of the programming outcomes (Alves et al., 2019) and the process to obtain them. We did this thanks to a varied set of variables that allow us to analyse general aspects (i.e., performance) and particular aspects (i.e., loops, conditionals, debugging).

Considering the development of CT a very broad objective, tools like Kodetu can contribute to answer questions such as 1) whether a certain challenge is appropriate for a specific group of learners, 2) if we are able to identify particularly difficult aspects in the development of CT and causes of drop–out (e.g., which challenges take longer, which levels require more attempts, etc. give us information about the difficult aspects in the game), 3) how should we design the interface and game dynamics according to the target age, and 4) whether limiting certain aspects (e.g., time available, number of blocks) to make learners use of certain CT resources (e.g., loops, alternatives, etc.) increases creativity and also difficulty.

We can see the potential that these tools have for the double objective of training and evaluating students. We propose to define associated control panels so teachers and researchers have access to the results of the students' activity, as the information they provide is rich and very valuable in the CT learning process.

RQ2. *How do personal features (gender, age, attitude towards technology, and prior programming knowledge) influence early CT experiences?*

We have analysed the influence of personal characteristics (gender, age, attitude towards technology, and previous programming knowledge) in the first CT experiences. The most appropriate age for this particular CT challenge starts at 11–12 years old, where we also observe a greater engagement than at 12–13 years. We interpret the especially good performance around age 11 in the psychological context surrounding the transition from primary to secondary school that happens in Spain. Splitting the data by school year reinforces this difference in school-level change. Pre–adolescent secondary school students may perceive Kodetu as being childish and they are probably much less motivated to play with it. Many factors could explain this superior performance at age 11–12, possibly due to the change in the academic stage. It is known that middle school students, aged around 11, are in the formal operational stage where they begin to think abstractly and are capable of deductive and logical reasoning (Wu, 2018).

Kodetu's programming challenges require not only TC knowledge but also spatial orientation skills. Multiple studies have found that 10–12 year olds still have problems with the correct use of the terms left and right (Murias et al., 2019; Rigal, 1994). This may suggest that age differences in our results could be related to this issue. However, the spatial vision requirements of the different levels remain constant throughout the sequence of levels used in Kodetu. Indeed, there are type A levels (sequences) that have exactly the same maze as type D levels (sequences + loops + conditionals) with significantly different results.

It seems essential to clearly specify which age is addressed and to provide differentiated tools and challenges for different ages (Alves et al., 2019). The performance linear progression before 11 years is in line with Seiter and Foreman (2013), who showed that the complexity of CT increases is consistent with the educational level, from 1st to 6th year of primary education.

Moreover, we should perform interviews and other qualitative research to better understand this finding.

Boys perform somewhat better than girls, though this tends to equalize in the more abstract or less mechanical challenges. Gender differences in attitudes toward game design and its impact on how females develop CT skills where also found in (Wu, 2018). The girls' preferred use of computers in a cooperative manner to achieve goals could explain this gender difference (Turkle, 1988).

Other variables seem not to be so relevant for the CT assessment. Students with prior programming knowledge perform better in all cases, and attitude towards technology is not a good predictor of performance, especially at the more complex levels. We should remark that these variables are collected through self-reported questions as in previous works (Kier et al., 2014), which may lead to a subjective interpretation of the term and self-response.

RQ3. *How the indicators that can be defined in a log-based system influence the performance of students in this particular type of system?*

In addition to the measures of the level reached and the time spent, we have generated, from data logged, a series of indicators that have shown useful to obtain better knowledge of the performance of students in this type of systems: number of attempts, the relation between code length and minimum length, changes made, dead code, etc. Furthermore, the study of the code originality, and the evolution of the code of each student, are relevant elements in the learning of CT. However, there are indicators that have shown not to be valuable to get insights on the development of CT, for example: previous knowledge of programming or the attitude towards technology. This may be due to the way these questions have been formulated, without specifying what we understood as programming or technology, and also because like in other works (Grover et al., 2016) they were self-reported.

Based on our results, we made a series of recommendations that should be considered when designing this type of system:

- The progression of difficulty should be tested by logging the behaviour of student groups in preliminary versions. We find similar proposals in the literature (Gujberova & Kalas, 2013).
- We should analyse the block-limitation per level. Except for the tutorial levels, we recommend a number of blocks small enough to avoid mechanical solutions and large enough to allow for the greatest expressiveness of code.

- To achieve the deepest interpretation and evaluation of the learning, it is convenient to integrate the training tool with analysis systems, with personalized indicators for the particular activity proposed (Grover et al., 2017a).
- The time stamping of logs should be studied very carefully and duplicated if necessary (i.e., browser time and database time). The times and thresholds that determine borderline cases are crucial in the analysis and interpretation based on interaction logs, and their editing requires a significant data-cleaning effort (Werner et al., 2013).
- It is highly relevant in the analysis of CT to incorporate tools that allow for simulation of the execution of the code built by the students to perform a more in-depth analysis of how learning happens. This kind of tool could support the monitoring, evaluation, and research to design better systems.

Despite the high number of participants and all results presented in the previous section, we are aware of the limitations of our work. The log-based quantitative methodology followed prevents us from having a complete knowledge of the development of CT. There is a need for complementary methods, such as observation and interviews, which can be combined and together comprise "systems of assessments" and provide a holistic view of student understanding (Grover, 2020), especially when regarding computational thinking practices and perspectives (Brennan & Resnick, 2012).

Nevertheless, automatically collected data allows us to make an exploratory analysis of the process and outcomes of participants and we got a first approach to the learning assessment of this type of tasks and platforms. On the other hand, the very fact of having such a high number of users and the set up of the sessions made it difficult to carry out a qualitative analysis that would have provided us with a deeper knowledge and understanding of some issues that remain open for future work.

Some limitations are related to the procedural aspect of the study, these are: 1) The set up of the sessions (i.e. little time available in the sessions, a key factor that led many students to leave early; sessions in the frame of a wider visit to the University, complemented with other workshops; it is difficult to interview students of these ages), 2) The way the test was carried out does not guarantee that external factors will not influence the game (i.e., classmates who comment on the solution of a level, students who are enthusiastic about their achievement and solve the challenge of their classmates), 3) the terms Programming and Technology in the pre-survey were not properly described and this could have led to misunderstand the terms and provide inconsistent answers to the self-informed questions. To overcome these limitations and continue this work, we suggest as future work to carry out a qualitative study that will provide us with deeper knowledge of the development of CT and to know the reasons for some of the findings derived from the data analysis.

Tools like Kodetu can play a key role supporting and delivering CT in primary and secondary education. This study contributes to the growing body of literature on CT tools and assessment, and -more importantly- to the still limited knowledge on log-based analysis of visual programming tools for children. Our findings show that students' data analysis can provide valuable insights on CT performance and learning progression, both for teachers and students. However, further research is needed to explore which factors and how affect students' performance in the acquisition of CT. Age and gender has shown to be variables that affect CT performance. Accordingly, teachers should design challenges that best fit the target students. Along with other assessment methods, Learning Analytics techniques may ease teachers' workload when dealing with large number of students and provide them with information hard to obtain manually. Showing individual information on performance to students and compared to the group in a comprehensive and usable way may give them feedback, enriching the formative assessment provided by other means. Finally, CT tools developers and instructional designers may use data analysis to better understand the interaction with the tool and the design of CT tasks. Future research may focus on the difficulty measures of CT tasks, drop-outs, and the integration of other CT concepts in this type of tools.

To facilitate the collaboration with other researchers interested in CT learning technologies, we released the complete dataset described here under an open access license: https://osf.io/x8dzp

Disclosure statement

No potential conflict of interest was reported by the author(s).

References

Alves, N. D. C., von Wangenheim, C. G., Hauck, J. C. R., & Borgatto, A. F. 2020. A large-scale evaluation of a Rubric for the automatic assessment of Algorithms And Programming Concepts. *Proceedings of the 51st ACM Technical Symposium on Computer Science Education* (pp. 556–562). Portland, US: ACM.

Alves, N. D. C., Von Wangenheim, C. G., & Hauck, J. C. R. (2019). Approaches to assess computational thinking competences based on code analysis in K-12 education: A systematic mapping study. *Informatics in Education, 18*(1), 17–39. https://doi.org/10.15388/infedu.2019.02

Baalsrud Hauge, J. M., Stanescu, I. A., Arnab, S., Moreno Ger, P., Lim, T., Serrano-Laguna, A., Lameras, P., Hendrix, M., Kiili, K., Ninaus, M., de Freitas, S., Mazzetti, A., Dahlbom, A., & Degano, C. (2015). Learning analytics architecture to scaffold learning experience through technology-based methods. *International Journal of Serious Games, 2*(1), 29-44. https://doi.org/10.17083/ijsg.v2i1.38

Baker, R. (2007). Modeling and understanding students' off-task behavior in intelligent tutoring systems. *Proceedings of the SIGCHI conference on Human factors in computing systems* (pp. 1059–1068). San Jose, CA, USA: ACM.

Barr, V., & Stephenson, C. (2011). Bringing computational thinking to K-12: What is Involved and what is the role of the computer science education community? *Acm Inroads, 2*(1), 48–54. https://doi.org/10.1145/1929887.1929905

Blikstein, P., Worsley, M., Piech, C., Sahami, M., Cooper, S., & Koller, D. (2014). Programming pluralism: Using learning analytics to detect patterns in the learning of computer programming. *Journal of the Learning Sciences, 23*(4), 561–599. https://doi.org/10.1080/10508406.2014.954750

Boe, B., Hill, C., Len, M., Dreschler, G., Conrad, P., & Franklin, D. (2013). Hairball: Lint-inspired static analysis of scratch projects. *Proceeding of the 44th ACM technical symposium on Computer science education*, 215–220. Denver, CO, USA.

Brennan, K., & Resnick, M. (2012). New frameworks for studying and assessing the development of computational thinking. *Proceedings of the 2012 annual meeting of the American Educational Research Association*, 1–25. Vancouver, British Columbia, Canada.

Buffum, P., Lobene, E., Frankosky, M., Boyer, K., Wiebe, E., & Lester, J. (2015). A practical guide to developing and validating computer science knowledge assessments with application to middle school. *Proceedings of the 46th ACM Technical Symposium on Computer Science Education*, 622–627. New York, NY, USA.

Cambruzzi, W. L., Rigo, S. J., & Barbosa, J. L. (2015). Dropout prediction and reduction in distance education courses with the learning analytics multitrail approach. *J. Ucs, 21*(1), 23–47. doi:10.3217/jucs-021-01-0023

Cartelli, A., Dagiene, V., & Futschek, G. (2010). Bebras contest and digital competence assessment: Analysis of frameworks. *International Journal of Digital Literacy and Digital Competence (IJDLDC), 1* (1), 24–39. https://doi.org/10.4018/jdldc.2010101902

Dasgupta, S., Hale, W., Monroy-Hernández, A., & Hill, B. (2016). Remixing as a pathway to computational thinking. (ACM, Ed.) *Proceedings of the 19th ACM Conference on Computer-Supported Cooperative Work & Social Computing (CSCW '16)*, 1438–1449. New York, NY, USA.

de Souza, A. A., Barcelos, T. S., Munoz, R., Villarroel, R., & Silva, L. A. (2019). Data mining framework to analyze the evolution of computational thinking skills in game building workshops. *IEEE Access, 7*, 82848–82866. https://doi.org/10.1109/ACCESS.2019.2924343

Denner, J., Werner, L., & Ortiz, E. (2012). Computer games created by middle school girls: Can they be used to measure understanding of computer science concepts? *Computers & Education, 58*(1), 240–249. https://doi.org/10.1016/j.compedu.2011.08.006

Denning, P. (2017). Remaining trouble spots with computational thinking. *Communications of the ACM, 60*(6), 33–39. https://doi.org/10.1145/2998438

Fields, D., Quirke, L., Amely, J., & Maughan, J. (2016). Combining big data and thick data analyses for understanding youth learning trajectories in a summer coding camp. *Proceedings of the 47th ACM Technical Symposium on Computing Science Education*, 150–155. New York, NY, USA.

Futschek, G. (2006). Algorithmic thinking: The key for understanding computer science. In *International conference on informatics in secondary schools-evolution and perspectives* (pp. 159-168). Springer, Berlin, Heidelber.

Grover, S. (2020). Designing an assessment for introductory programming concepts in middle school computer science. *Proceedings of the 51st ACM Technical Symposium on Computer Science Education* (pp. 678–684). SIGCSE '20. New York, NY, USA: Association for Computing Machinery. https://doi.org/10.1145/3328778.3366896.

Grover, S., Basu, S., Bienkowski, M., Eagle, M., Diana, N., & Stamper, J. (2017). A framework for using hypothesis-driven approaches to support data-driven learning analytics in measuring computational thinking in block-based programming environments. *ACM Transactions on Computing Education (TOCE)*, *17*(3), 1–25. https://doi.org/10.1145/3105910

Grover, S., Bienkowski, M., Niekrasz, J., & Hauswirth, M. (2016). Assessing problem-solving process at scale. *Proceedings of the Third ACM Conference on Learning @Scale* (pp. 245–248). L@S '16. New York, NY, USA: Association for Computing Machinery. https://doi.org/10.1145/2876034.2893425.

Grover, S., Cooper, S., & Pea, R. (2014). Assessing computational learning in K-12. *Proceedings of the 2014 conference on Innovation & technology in computer science education* (pp. 57–62). Uppsala, Sweden.

Grover, S., & Pea, R. (2013). Computational thinking in K–12: A review of the state of the field. *Educational Researcher*, *42*(1), 38–43. https://doi.org/10.3102/0013189X12463051

Gujberova, M., & Kalas, I. (2013). Designing productive gradations of tasks in primary programming education. *Proceedings of the 8th Workshop in Primary and Secondary Computing Education (WiPSE '13)* (pp. 108–117). New York, NY, USA: ACM.

Hubwieser, P., & Mühling, A. (2014). Playing PISA with bebras. *Proceedings of the 9th Workshop in Primary and Secondary Computing Education* (pp. 128–129). Berlin, Germany.

Israel-Fishelson, R., & Hershkovitz, A. (2020). Shooting for the stars: Micro-Persistence of students in game-based learning environments.In *Early Warning Systems and Targeted Interventions for Student Success in Online Courses* (pp. 239-258). IGI Global. https://doi.org/10.4018/978-1-7998-5074-8.ch012

Kelleher, C. (2006). *Motivating Programming: Using storytelling to make computer programming attractive to middle school girls* (No. CMU-CS-06-171). Carnegie-Mellon Univ.

Kier, M. W., Blanchard, M. R., Osborne, J. W., & Albert, J. L. (2014). The development of the STEM Career Interest Survey (STEM-CIS). *Research in Science Education*, *44*(3), 461–481. https://doi.org/10.1007/s11165-013-9389-3

Kong, S.-C. (2019). Components and methods of evaluating computational thinking for fostering creative problem-solvers in Senior Primary School Education. In S.-C. Kong & H. Abelson (Eds.), *Computational thinking education* (pp. 119–141). Springer Singapore. https://doi.org/10.1007/978-981-13-6528-7_8

Lewis, C. (2011). Is pair programming more effective than other forms of collaboration for young students? *Computer Science Education*, *21*(2), 105–134. https://doi.org/10.1080/08993408.2011.579805

Lye, S. Y., & Koh, J. H. L. (2014). Review on teaching and learning of computational thinking through programming: What is next for K-12? *Computers in Human Behavior*, *41*, 51–61. https://doi.org/10.1016/j.chb.2014.09.012

Maloney, J. H., Peppler, K., Kafai, Y., Resnick, M., & Rusk, N. (2008). Programming by choice: Urban youth learning programming with scratch. *Proceedings of the 39th SIGCSE Technical Symposium on Computer Science Education*, *40*(1), 367–371. Portland, OR, USA.

Meerbaum-Salant, O., Armoni, M., & Ben-Ari, M. (2011). Habits of programming in scratch. *Proceedings of the 16th annual joint conference on Innovation and technology in computer science education*, 168–172. Darmstadt, Germany.

Meerbaum-Salant, O., Armoni, M., & Ben-Ari, M. (2013). Learning computer science concepts with scratch. *Computer Science Education*, *23*(3), 239–264. https://doi.org/10.1080/08993408.2013.832022

Moreno-León, J., & Robles, G. (2015). Dr. Scratch: A web tool to Automatically Evaluate Scratch Projects. (ACM, Ed.) *Proceedings of the Workshop in Primary and Secondary Computing Education (WiPSCE '15)*, 132–133. Virtual Event, Germany. https://doi.org/10.1145/2818314.2818338

Mühling, A., Ruf, A., & Hubwieser, P. (2015). Design and first results of a psychometric test for measuring basic programming abilities. *Proceedings of the workshop in primary and secondary computing education*, 2–10. London, UK.

Murias, K., Slone, E., Tariq, S., & Iaria, G. (2019). Development of spatial orientation skills: An fMRI study. *Brain Imaging and Behavior*, *13*(6), 1590–1601. https://doi.org/10.1007/s11682-018-0028-5

National Research Council. (2010) . *Report of a workshop on the scope and nature of computational thinking*. National Academies Press.

NRC-National Research Council. (2012). *A Framework for K-12 science education: Practices, crosscutting concepts, and core ideas*. The National Academies Press.

Papert, S. (1980). *Mindstorms: Children, computers, and powerful ideas*. Basic Books, Inc.

Pelánek, R., & Effenberger, T. (2020). Design and analysis of microworlds and puzzles for block-based programming. *Computer Science Education*, 1–39. https://www.tandfonline.com/doi/abs/10.1080/08993408.2020.1832813

Piech, C., Sahami, M., Koller, D., Cooper, S., & Blikstein, P. (2012). Modeling how students learn to program. *Proceedings of the 43rd ACM technical symposium on Computer Science Education*, 153–160. Raleigh, North Carolina, USA.

Rigal, R. (1994). Right-left orientation: Development of correct use of right and left terms. *Perceptual and Motor Skills, 79*(3), 1259–1278. https://doi.org/10.2466/pms.1994.79.3.1259

Robins, A., Rountree, J., & Rountree, N. (2003). Learning and teaching programming: A review and discussion. *Computer Science Education, 13*(2), 137–172. https://doi.org/10.1076/csed.13.2.137.14200

Román-González, M., Pérez-González, J., & Jiménez-Fernandez, C. (2017). Which cognitive abilities underlie computational thinking? Criterion validity of the computational thinking test. *Computers in Human Behavior, 72*C (July 2017), 678–691. https://doi.org/10.1016/j.chb.2016.08.047

Ruipérez-Valiente, J. A., Muñoz-Merino, P. J., Leony, D., & Kloos, C. D. (2015). ALAS-KA: A learning analytics extension for better understanding the learning process in the Khan Academy platform. *Computers in Human Behavior, 47*, 139–148. https://doi.org/10.1016/j.chb.2014.07.002

Sadler, D. R. (1989). Formative assessment and the design of instructional systems. *Instructional Science, 18*(2), 119–144. https://doi.org/10.1007/BF00117714

Seiter, L., & Foreman, B. (2013). Modeling the learning progressions of computational thinking of primary grade students. *Proceedings of the ninth annual international ACM conference on International computing education research*, 59–66. San Diego, California, USA.

Selby, C., & Woollard, J. (2013). Computational thinking: The developing definition. *19th Annual Conference on Innovation and Technology in Computer Science Education* Canterbury, Great Britain.

Sorva, J. (2012). *Visual program simulation in introductory programming education*. Aalto University.

Spacco, J., Denny, P., Richards, B., Babcock, D., Hovemeyer, D., Moscola, J., & Duvall, R. (2015). Analyzing student work patterns using programming exercise data. *Proceedings of the 46th ACM Technical Symposium on Computer Science Education*, 18–23. New York, NY, USA.

Syslo, M. (2015). From algorithmic to computational thinking: On the way for computing for all students. *Paper presented at the proceedings of the 2015 ACM Conference on Innovation and Technology in Computer Science Education* Vilnius, Lithuania. https://doi.org/10.1145/2729094.2742582

Tang, X., Yin, Y., Lin, Q., Hadad, R., & Zhai, X. (2020). Assessing computational thinking: A systematic review of empirical studies. *Computers & Education, 148*(2020), 103798. https://doi.org/10.1016/j.compedu.2019.103798

Tedre, M., & Denning, P. (2016). The long quest for computational thinking. *Proceedings of the 16th Koli Calling International Conference on Computing Education Research (Koli Calling '16)* (págs. 120–129). New York, NY: ACM. https://doi.org/10.1145/2999541.2999542

Tew, A. E., & Guzdial, M. (2011). The FCS1: A language independent assessment of CS1 knowledge. *Proceedings of the 42nd ACM technical symposium on Computer science education*, 111–116. New York, NY, USA.

Torrance, H. (1995). *Evaluating authentic assessment: Problems and possibilities in new approaches to assessment*. Open University Press.

Touretzky, D., Gardner-mccune, C., & Aggarwal, A. (2017). Semantic reasoning in young programmers. (ACM, Ed.). *Proceedings of the 2017 ACM SIGCSE Technical Symposium on Computer Science Education (SIGCSE '17)*, 585–590. Seattle, Washington, USA. https://doi.org/10.1145/3017680.3017

Turkle, S. (1988). Computational reticence: Why women fear the intimate machine. In C. Kramarae (Ed.), *Technology and women's voices: Keeping in touch*(p. 41-61). Routledge.

Vihavainen, A., Luukkainen, M., & Ihantola, P. (2014). Analysis of source code snapshot granularity levels. *Proceedings of the 15th Annual Conference on Information technology education* (págs. 21–26). Atlanta, Georgia, USA: ACM.

Werner, L., Denner, J., Campe, S., & Kawamoto, D. (2012). The fairy performance assessment: Measuring computational thinking in middle school. *Proceedings of the 43rd ACM technical symposium on Computer Science Education*, 215–220. New York, NY, USA.

Werner, L., McDowell, C., & Denner, J. (2013). Middle school students using Alice: What can we learn from logging data? *Proceeding of the 44th ACM technical symposium on Computer science education*, 507–512. Denver, Colorado, USA.

Wing, J. (2006). Computational thinking. *Communications of the ACM*, *49*(2), 33–35. https://doi.org/10.1145/1118178.1118215

Wing, J., & Stanzione, D. (2016). Progress in computational thinking, and expanding the HPC community. *Communications of the ACM*, *59*(7), 10–11. https://doi.org/10.1145/2933410

Wolz, U., Hallberg, C., & Taylor, B. (2011). Scrape: A tool for visualizing the code of Scratch programs. *Poster presented at the 42nd ACM Technical Symposium on Computer Science Education* Dallas, TX.

World Economic Forum. (2015). *New vision for education: Unlocking the potential of technology*. Vancouver, BC: British Columbia Teachers' Federation. *http://www3.weforum.org/docs/WEFUSA_NewVisionforEducation_Report2015.pdf*

Wu, M. L. (2018). «Educational game design as gateway for operationalizing computational thinking skills among Middle School Students». *International Education Studies*, *11*(4), 15. 29 de marzo de 2018. https://doi.org/10.5539/ies.v11n4p15

Zur-Bargury, I., Pârv, B., & Lanzberg, D. (2013). A nationwide exam as a tool for improving a new curriculum. *Proceedings of the 18th ACM conference on Innovation and technology in computer science education*, 267–272. Canterbury, England, UK.

Assessing computational thinking in libraries

David Weintrop, Shandra Morehouse and Mega Subramaniam

ABSTRACT

Background and Context: Computational thinking (CT) is an essential skill for all youth to succeed in our technology and information-rich world. While CT has a growing presence within K-12 classrooms, libraries play an essential role in introducing these critical skills to all.

Objective: Assessing learning in libraries is difficult given their informal nature. This is especially true when trying to assess a new and ill-defined construct like CT. A first step towards better supporting informal educators is to identify the motivations for bringing CT into informal spaces and to understand the current state of assessing CT in public libraries.

Method: Thirty-seven interviews with library staff from across the United States were conducted and analyzed.

Findings: This work reveals the variety of motivations justifying the inclusion of CT programming for youth in libraries, the challenges associated with assessing CT in libraries, and identifies the assessments library staff would like to be able to conduct for their own CT-related programming.

Implications: This work advances our understanding of the current state of CT assessment in public libraries and lays the groundwork for future work seeking to meet the needs of those tasked with bringing CT to youth beyond the classroom.

Introduction

As the presence and importance of technology expands, it is essential that all children growing up in an increasingly digital landscape have a foundational understanding of computing. This includes not just how to use computational devices, but an understanding of what computing can do and how to express ideas in computationally meaningful ways. Often captured under the term *Computational Thinking* (CT), these ideas include broadly useful concepts and practices from the field of computer science including problem decomposition, abstraction, algorithms, pattern recognition, and concepts related to computer programming (Curzon, 2019; Grover & Pea, 2013; Wing, 2006). While CT has a growing presence in K-12 classrooms, informal spaces such as public libraries, museums, and youth community centers also play important roles in providing opportunities for children to develop these essential 21st-century skills.

The role of informal spaces, like public libraries, in introducing youth to CT stems from the flexibility and openness afforded by such settings. Where schools and teachers are constrained by rigid schedules and externally defined learning outcomes, informal spaces such as public libraries have the freedom to provide more open-ended learning experiences and can incorporate more novel technologies and innovative pedagogical approaches (Subramaniam et al., 2018a). Public libraries are well-suited for CT learning activities that draw on novel technologies (e.g., robotics toolkits, emerging digital media) (Garmer, 2014) and play an essential role in providing physical access to such technologies (Davis et al., 2018) and offering intellectual support in the form of CT programming to underserved youth in their communities (Braun & Visser, 2017). Additionally, schools often limit potential CT learning experiences to a subset of students or are unable to provide such experiences at all (Wang et al., 2016) whereas libraries serve all in their communities[1].

While libraries share characteristics of other informal learning contexts like museums, after school clubs, and summer camps, libraries play a critical role in bringing CT to all learners given their ubiquity, accessibility, and the role they already play in supporting communities. In the United States, people visit public libraries more often than they go to the movies (1.35 billion visits vs. 1.25 billion admissions) (American Library Association, 2019), with most public libraries (76.52%) serving a population area of fewer than 25,000 people (IMLS, 2019). This situates libraries at the forefront of innovation and experimentation when it comes to creating effective and engaging ways to introduce youth to CT as in the United States, "the library is the only place that is truly accessible and inclusive for all" (American Library Association, 2019, p. 6).

At the same time, there are challenges that accompany bringing CT into informal contexts. Central among them is the question of assessment. While the goals and learning outcomes of instructional experiences in libraries are distinct from those of classrooms, it is important for librarians to understand if and how their CT programming is succeeding. The challenge of conducting assessments in public libraries is compounded by the variety of formats of CT learning experiences offered, the "drop-in" nature of these experiences, duration of these experiences, and the unavailability of appropriate tools to conduct such assessment. This paper seeks to explore the challenges of assessing CT in libraries. Given the importance of all youth developing foundational CT skills and the potential that public libraries play in introducing these ideas, it is essential that we understand how to take full advantage of libraries as part of the larger goal of bringing CT to all. To make progress on this goal, we pursued the following set of research questions:

Why do public library staff seek to bring computational thinking into their libraries? What are their goals for the computational thinking programs they offer?

What challenges do libraries face when assessing computational thinking?

How would public library staff like to assess computational thinking?

To answer these questions, we interviewed public library staff serving youth from across the United States, asking them about their goals and experiences in conducting CT programs in their libraries, the challenges they encounter when assessing CT, and any aspirations they might have for CT assessment. These interviews included library staff from rural, suburban, and urban contexts.

This work advances our understanding of why libraries are seeking to incorporate CT in libraries and their potential for innovative and impactful CT programming. We gain insight into the challenges associated with assessing the CT experiences that currently happen in libraries. In addition to laying the groundwork for future work that support those bringing CT to youth in libraries, this work sheds light on the gaps that exist within the currently used CT assessment tools and allow us to explore other modes of assessments to help libraries reach their full potential as powerful venues for helping all youth have positive CT learning experiences.

This paper continues with a review of the literature this work builds upon. Next, the paper presents the method used for this study followed by the findings section, broken into three sections that align with the three driving research questions for this work. The paper concludes with a discussion of implications, limitations, and future work. Specifically, the discussion looks across the result sections identifying ways in which goals and assessments do and do not align, highlighting the need for future work on CT assessment instruments to best support library staff in fulfilling their critical role in bringing CT to a wide swath of the youth population. Collectively, the contribution of this work is an understanding of why libraries are seeking to incorporate CT into their youth programming, the challenges they face in doing so, and insight into how they hope to evaluate the CT programming they offer.

Literature review

In this section, we begin by explaining our focus on libraries as contexts where CT learning occurs and then review the two bodies of work that this work draws most directly from and that we seek to contribute to. First, we discuss prior research bringing CT into library contexts followed by a review of the state of CT assessments with an emphasis on efforts to design and conduct assessments of CT in informal settings.

Libraries as contexts for learning

Public libraries have evolved to meet the needs of their communities, including the K-12 populations. While customary services such as book checkouts and reader advisory, and legacy programs such as story times and book clubs continue, public libraries have evolved "to distinctive and purposeful programming, resources, and services, both inside the library and out . . . " (American Library Association, 2019, p. 19) serving diverse needs in their communities, including contribution to cultivating competencies and interest that are necessary for pursuing careers in computing (Subramaniam et al., 2018b). Importantly, nearly all public libraries offer access to computers and free wifi for all visitors (American Library Association, 2019).

Public libraries play a significant role in supporting communities. In the United States, 96% of the population lives within a public library legal service area (American Library Association, 2019), indicating that public libraries are institutions that can are physically accessible to a vast majority of communities. Globally, more there are more than 380,000 libraries with internet access than 430,000 public and community libraries serve 132 countries around the world (IFLA Library Map of the World, 2020). These libraries serve 551.3 million users around the world with 4,376.9 million total visits (IFLA). Libraries are particularly important as resources for materials and technological resources in rural areas and also provide a significant amount of programming for youth (Davis et al., 2018; Phillips et al., 2019). In 2016, public libraries offered 503,334 programs for young adults (ages 12 to 18) and 2.85 million programs for children ages 11 and under (IMLS, 2019). Further, the more rural the library is, the more children and young adults' programs are offered (IMLS, 2019). With 113 million attendees to public library programs in 2016, (American Library Association, 2019) libraries are in a position to provide access and computer-based programming to youth who may not have access to the technology at home or in their school (Martin, 2017).

While public libraries have been offering technology-enabled programs for youth for more than a decade (Braun et al., 2014; Lee, 2019; Subramaniam et al., 2018a), libraries are increasingly offering programs tailored to the development of interest and competencies in STEM, with 76% of the state library agency chief officers stating they either are planning on implementing or have ongoing STEM programming efforts at their state (Rose, 2015).

Many of these programs that intend to develop competencies and interests in STEM leverage design principles of the connected learning framework (Hoffman et al., 2016; Subramaniam et al., 2018a) to design their programs to be authentic to youth, interest-driven, fun, and not be like "school". Increasingly, this programming includes concepts or practices associated with computational thinking (Subramaniam et al., 2019).

Computational thinking in libraries

Although there are many definitions of CT, in this paper, we define CT as a set of concepts and practices from the field of computer science that are useful when using computers and technology to solve problems. This includes problem decomposition, developing and using abstractions, debugging, defining algorithms, and concepts grounded in the practice of programming such as iteration, parallelization, and conditional logic (Curzon, 2019; Grover & Pea, 2013). Wing (2006) states that "computational thinking is a fundamental skill for everyone, not just the computer scientists" (p. 3) and argues that CT be taught along with other essential skills such as reading and writing.

As discussed above, public libraries have become a hub for informal CT learning in communities. As facilitators of lifelong and informal learning, libraries have become major players in promoting digital literacy for youth in their community (American Library Association, 2020; Taylor et al., 2018). As the need for CT education expands, libraries have found ways to include CT into their programming. For example, the American Library Association (ALA) partnered with Google for the Libraries Ready to Code (RtC) initiative and provided grants to 250 libraries to implement coding and CT programs (Braun & Visser, 2017). In addition to grant initiatives like RtC, CT is being included library training programs (Taylor et al., 2018) and continuing education efforts like the Train the Trainer program (Young Adult Library Services Association, 2018), and the Future Ready with the Libraries program (Young Adult Library Services Association, 2016) offered by the Young Adult Library Services Association (YALSA).

Wing (2006) argues that CT is for everyone and that it is not limited to computer science but has everyday applications. Many librarians build on this idea when it comes to planning, advertising, and facilitating their CT events. Much of the library staff that plans and facilitates CT programming are new to the concepts involved; therefore, when deciding how to facilitate these programs or start a code club, many library staff look to low cost and easy to use tools and activities that they can use in their library (Braun & Visser, 2017). Some library programs focus on teaching programming skills using block coding programs like Scratch (Braun & Visser, 2017). Others facilitate the learning of CT concepts using various tools (Guidara, 2018; Lee & Recker, 2018; Prato, 2017; Wing & Meyers, 2014). For example, robotics is a common way to facilitate CT education in libraries. Products like Dash Robots, Ozobots, and Lego Mindstorms are used in CT programs and provide an easy entry for library staff who have little to no experience in CT skills and education (Prato, 2017). Library staff also use tools like Raspberry Pi and paper circuits as an easy way to teach computational skills without requiring previous knowledge of CT, and in the context of paper circuits, a low-cost alternative to other tools (Lee & Recker, 2018; Wing & Meyers, 2014). Similarly, many libraries offer "unplugged" activities, using pen and paper as another low cost, low stakes alternative to these tools. (Guidara, 2018; Prato, 2017). Library staff also use ready-made curricula, such as Google's CS First and Girls Who Code to facilitate their programs (Braun & Visser, 2017).

Despite the current efforts across the US to bring CT to libraries, we do not yet have a clear picture of all the different types of programming that is being held, nor do we fully understand their motivations and goals in hosting these programs. In addition, CT is still a relatively new term for the library profession, which affects the adoption of CT programming by library staff and their ability to facilitate CT learning.

Assessing computational thinking

In this section, we review the *what* and the *how* of assessing CT. In their review of existing CT assessments, Tang et al. (2020) found that "many studies chose to measure programming or computing concepts as representations of CT skills" (p. 18). Brennan and Resnick (2012) identify three areas of CT to assess: computational concepts (i.e. sequences, loops, conditionals), computational practices (i.e. testing and debugging, reusing and remixing), and computational perspectives (i.e. observing shifts in perspective). There are many different approaches and tools used to assess CT. Some examples include selected-response tests (e.g., Grover et al., 2015; Weintrop et al., 2014), portfolio assessment (e.g., Brennan & Resnick, 2012); questionnaires or surveys (Mouza et al., 2016), interviews (e.g., Berland & Lee, 2011), design scenarios (Brennan & Resnick, 2012), and pre- and post- tests (Arastoopour Irgens et al., 2020). CT has also been assessed through less tangible means, for example, Lee and Recker (2018) assess their paper circuits program by having participants share their projects with the group, or through informal competitions and exhibits.

Shifting out focus to informal spaces, like after school programs, workshops, and summer programs, the research shows a mix of formal and informal assessment tools. These informal CT programs often used formal assessments, like worksheets, to measure a participant's knowledge and understanding of CT concepts. (e.g., Mouza et al., 2016; Pugnali et al., 2017; Wu, 2018). Pugnali et al. (2017) assessed CT skills based on the Positive

Technical Development Framework that is focused on six aspects of CT engagement: Communication, Collaboration, Community Building, Content Creation, Creativity, and Choices of Conduct. The use of formal assessments is often coupled with other forms of assessment like portfolios of participants' projects and observations (Mouza et al., 2016; Wu, 2018). Informal learning spaces also employ non-formal assessment tools to assess learning and the attitudes of youth who participate in their programs. For example, Denner et al. (2012) assessed the portfolios of video games created by their participants. Other work in informal learning spaces has assessed participants' knowledge and attitudes through participant reflections, including the work of Thomas et al. (2017) who reviewed their participant's daily reflective journal to gain more insight into the attitudes and thoughts of their participants.

Assessment in libraries can be a challenge as the library is an informal learning space that is not well suited for traditional modes of assessment, even less so than other informal spaces which may be more structured in format. However, this does not mean there are not efforts to assess library programs. One example of work seeking to assess library programs is the Public Library Association's (PLA) Project Outcomes. Project Outcomes is a toolkit designed to help libraries measure their impact in a community. The toolkit provides surveys, a survey management tool, and easy to use ways to measure and analyze outcomes. The toolkit focuses on seven different areas of library service including community engagement, digital learning, education/lifelong learning, early childhood literacy, and job skills (PLA). Traditionally, attendance is a primary means of assessing library programs (American Library Association, 2020; Cole et al., 2013; Garmer, 2014). However, within the library field, there is a big push to measure the success of a program beyond looking at the numbers. Cole et al. (2013) refer to this as outcomes-based assessment. In their example, Cole et al. (2013) establish outcomes for their summer reading program and then use surveys at the end of the program to evaluate how/if these outcomes were met.

Since libraries are informal learning spaces, assessments also tend to be informal; mainly with anecdotal evidence and verbal feedback. In their work on capturing connected learning in libraries, the Connected Learning Alliance uses talkback boards to capture learning in libraries. Talkback boards are poster-sized displays that provide a few statements (i.e. *"Today I discovered an interest or talent that I didn't know I had"*) that participants use a sticker to indicate the statement they most agree with (Penuel et al., 2019). These statements serve to capture the desired outcomes of the program (Penuel et al., 2019). Another way libraries assess their programs is by using verbal feedback and surveys, like in Wing and Meyers (2014) Raspberry Pi programs.

Many assessments for CT in K-12 involve types of formal assessments like testing and surveys that are not conducive for informal spaces like libraries. With the limitations of library environments, the challenge then becomes how do library staff know that their participants are learning and that they are meeting their goals for CT programming. While there is research on how to get CT programming into libraries, CT assessment in libraries is either one small part of a bigger study, or not discussed. By focusing on CT assessment in public libraries, this paper seeks to understand the current CT assessment practices in libraries, the challenges that come with assessing CT in libraries, and what library staff would like to assess in their CT programs.

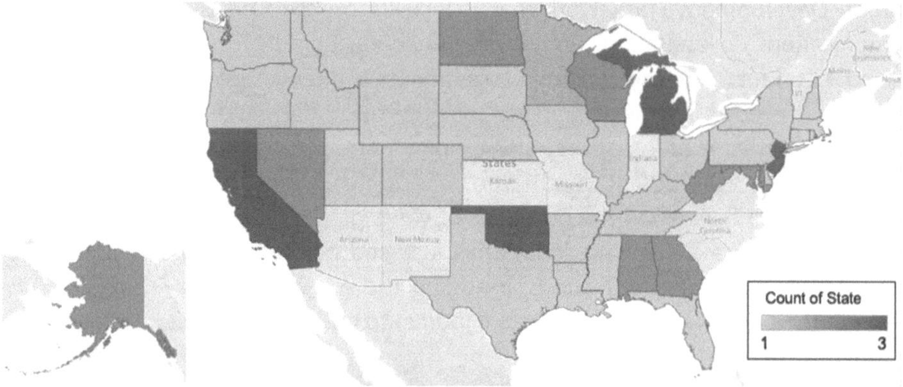

Figure 1. Distribution of participants by state.

Methods

To better understand the current landscape of CT activities in libraries, we interviewed 37 youth-serving library staff from across the United States (US) (Figure 1). Participants were recruited mainly through the mailing lists of ALA and YALSA, from the professional networks of the research team, and project partners. We collected information about the participants, such as their position, their location, population served (self-identified as rural, suburban, urban), and how many years they have been in their position as part of the sign up process. A summary of participant characteristics can be found in Table 1.

We conducted semi-structured interviews via phone or video conferencing software. These virtual interviews were recorded and lasted between 30–60 minutes. Interviews served as the primary methodology given the intent to obtain a national view of CT in libraries. Semi-structured interviews are well suited to capture the voices of library staff and their experiences and challenges in offering CT programs in libraries which are the focus of this research. The interview protocol covered topics such as the interviewee's experience with CT and computer science, the programs they ran at their library that included CT, the populations they worked with, the challenges they faced, and how they currently measured the success of their programs. Interviews were then de-identified and

Table 1. Demographic information for the 37 study participants.

Population Density	# of Participants	Average # of Years Experience	Job Title
Urban	8	13 years	Youth/Teen Librarian (4) IT/Technology Librarian (3) Branch Manager (1)
Suburban	14	7.8 years	General Librarian (5) Youth/Teen Librarian (5) IT/Technology Librarian (3) Branch Manager (1)
Rural	15	6.6 years	Youth/Teen Librarian (9) Branch Manager/Director (3) General Librarian (2) IT/Technology Librarian (1)

transcribed. The full protocol is available as an online supplement. Participants signed consent forms before their interview date and were offered a 25 USD Amazon gift card in return for their participation. We obtained institutional approval for human subject research prior to the start of data collection.

We employed thematic analysis (Boyatzis, 1998) to create a coding scheme aligned with our research questions. This methodological approach was used as we wanted to better understand the state of CT and how it is being assessed as experienced by those who are actually designing and running CT programming within libraries, rather than bringing a particular framework from the academic literature to shape and organize responses. To create a codebook, the research team individually coded a transcript and worked together to create an initial codebook. We then used this codebook to code an additional two transcripts individually, then met to discuss the results of our coding, making adjustments and clarifying code definitions as needed. The final codebook contained 16 codes that captured topics such as library demographics, resources used in CT activities in the library, the goals of CT activities, the challenges the library staff faced running these activities, and the assessment tools used for these activities. The codebook was then applied to all 37 interviews by one member of the team with an additional team member reviewing and modifying the coded transcripts as needed. Disagreements in coding were noted in memos in the analysis software and were later discussed and resolved between researchers (Smagorinsky, 2008).

To address the research questions in this paper, we specifically focus on four codes that emerged in our analysis: CT Assessment, Aspirational CT Assessment, Challenges, and CT Motivations/Goals/Outcomes. CT Assessment and Aspirational CT Assessment were used to capture participant's current efforts in assessment, their thoughts on assessing their programs, and the ways they wish they were able to assess their programs. The Challenges code captures a variety of challenges librarians identified when it came to running CT programming including staffing problems, staff knowledge, funding, infrastructure, and challenges with assessment. Finally, CT Motivations/Goals/Outcomes captures the various motivations librarians have for their CT programs, the goals they hope to accomplish, and the outcomes they would like to see in the youth participants.

Findings

The findings section is broken down into three sections, each of which answers one of our driving research questions. First, we present motivations for libraries to incorporate CT into the programs that they offer. Next, we present the challenges library staff encounter in trying to assess the CT programming at their libraries and then we conclude by presenting how library staff aspire to assess CT.

Motivation for bringing computational thinking into libraries

Why do library staff seek to bring CT into their libraries? What do they hope to accomplish in doing so?

As part of the interview protocol, we asked participants why they thought learning CT was important for youth, the goals they had for their programs, and the skills and attitudes they hoped to develop as a result of their programs. We asked these questions

Table 2. Themes from library staff responses asking about motivations for bringing CT into their libraries.

Code	Sample Quote
Positive CT Experiences	*"I want them to walk away like knowing that they can do this. It's not just for like certain people, it's not just for people who are like really into science or really into engineering, like anyone can do it."*(Andrea[2] – rural, Midwest)
Conceptual Learning	*"I'm wanting them to develop basic coding skills."* (Brittany – rural Southwest)
Career Preparation	*"The goal really for me for our program is for youth in [our community] to have a good future"* (Cait – urban, West coast)
Relating CT Programming to Schools	*"We felt we'd be able to offer the kid's something that they are not getting at school, and the parents were really interested in it"* (Brad – suburban, Northeast)
Access to Technology	*"My first goal is just to make something available to kids who maybe would not normally have access to it"* (Sandra – rural, Midwest)
Foster Collaboration	*"I will only get six [robots] out for a group of 12 [kids] because they have to share because that's part of working as a team and that is something that's a life skill."* (Molly – suburban, Southeast)

to better understand the motivations behind bringing CT into libraries. In reviewing the responses to these questions, we found justifications that aligned with the larger literature around CT as well as some library-specific goals for their CT programming. Table 2 provides a sample quote for each theme identified.

Looking across the responses, some motivations cited are not unique to CT, but common to many types of library programming, such as the desire for positive experiences with the content of the programming or to prepare youth for their futures. In other places, the motivations have particular characteristics that uniquely intersect with CT, such as future career preparation and access to technology that is essential for CT learning experiences. We present motivations that are common across library programming as well as those unique to CT to present the fullest picture of why library staff are working to bring CT into their buildings.

The most common reason our interviewees cited for introducing CT into the programming offered at their libraries was to allow youth to have positive early CT experiences and develop favorable perceptions and attitudes towards CT. For example, Tyler, from a suburban library in Northeastern US responded: *"Well, first and foremost, I just want to make it fun for the kids so that they have a good time and a good experience with engineering and computer programming concepts and that they feel engaged and that they feel like they're learning."* Other responses were related to specific attitudinal or dispositional goals. For example, several library staff referenced introducing CT as a means to help develop desirable dispositions such as perseverance, *"Definitely them learning that mistakes are good. That's been something that has been a real focus for us. To keep going, to persevere, that mistakes aren't bad, that that's how we learn"* (Jenna, rural library in the Northeastern US). Confidence was another frequently cited attitudinal outcome. When asked what skills or attitudes she hoped to develop, Cait, who works in an urban library in the Western US responded with the single word *"confidence"*, when asked to elaborate she continued *"Confidence in, in this new technology, in technology and thinking in a computational way."*

A second cited reason for bringing CT into the library was to help youth learn foundational ideas related to computing and technology. Along with the quotes in Table 1 demonstrating the goal of conceptual understanding, library staff also focused on the relationship between the technology youth see and use every day and the learning goals of their CT programming. "*The kids are using this technology every day. It'd be great if they knew what went into what they're using*" (*Brad, suburban Northeast*). Similar motivations can be seen in other comments focusing on digital media broadly rather than the technological aspects: "*not only do I want kids to understand how media is created, so they can understand the messages that are coming from that media. I want them to be able to evaluate it, analyze it ... In order to do that, they need to know how media works whether it's a book or some digital tool. And one of the best ways for kids to better understand how those tools work is to be able to create using that tool. And that ultimately allows them also to communicate with that tool*" (*Caroline, rural Western US*). These motivations are interesting as they closely relate to communication and information dissemination, objectives central to the missions of libraries as resources for creating an informed citizenry.

The economic potential of CT-related jobs was also raised as a motivation for programming. For example, Phyllis, a librarian from an urban library in the Southeastern US, stated a goal of her CT programming was to "*impact [young people] so they have a brighter future and so they will consider computer science because there's a lot of money in it*". Another variant of this goal was a broader motivation to prepare youth for the increasingly technological landscape irrespective of future profession. This can be seen in this response from Alice, an urban librarian from the Midwest, who said: "*the big goal would be to have all students in our county study computational science in elementary school so that they have, that they are better prepared for the future.*" Here we see the goal of CT extend beyond professional preparation to the larger goal of being prepared for whatever shape our technological future may take.

Another goal for CT programming in libraries was framed around the relationship between libraries and schools. In some cases, the motivation was to prepare youth who come to their CT programming for future computing classes in their schools while Molly, a suburban librarian from the Southeastern US, expressed this in her response: "*We want them to be prepared before they get to high school ... so [youth are] already exposed to it before they are exposed in the classroom*". Other times, library staff viewed their CT programming as serving a complementary role to that of schools. Kimberly, a rural librarian from the Western US, stated: "*we're reinforcing what they're getting from the schools and schools are reinforcing what they're getting at the library*". This sentiment of library programs supporting and supplementing schools was also raised in relation to contexts where local schools do not offer computing-related courses. "*The initial reason we really wanted to offer these programs are because the school system in our town doesn't have any coding classes for kids. So they don't have a computer science class.*" (Brad – suburban, Northeast). It is important to note that library staff did not view their goal as replacing schools or the role they play, Brad pointed this out in saying: "*we're not going to be able to teach them something that a school can teach .... [our CT program] is only once a month*", but rather librarians saw their role as helping youth in the community engage with the powerful ideas of CT and to provide access to tools to let them do so.

In discussing the relationship with schools, library staff highlight unique affordances of libraries that make them distinct from conventional classroom contexts. For example, the fact that learning in a library is not formally assessed was viewed as a feature that would help kids explore new ideas they might be hesitant to in a school-based context: "*the goals right now are to introduce CT concepts to kids in really stress-free ways or low-stress ways. They're not being tested. So that and exposed them to different kinds of projects and tools that they might be interested in learning more about."* (Caroline, rural Western US). The view was echoed by Nicole (suburban, Western US) who said *"[youth] don't need another classroom setting especially if they're moving into middle school and high school where they have the rotating classes, rotating teachers, all of the things that they're learning. They have to sit still. They have to pay attention, and that's not what the library is. We don't need to fill that in their lives. They already have enough of that."* These quotes highlight the distinct and complementary role of libraries as contexts for learning CT.

A final noteworthy motivation from these interviews was in the way library staff viewed their own institutions as a community resource when it came to the topic of CT and related technologies, specifically in response to issues of access to tools and technology. *"I think the main goal is still access. We want to give access to technology that is expensive and difficult to obtain for lower-income families"* (Nicole – suburban, West coast). This focus on access for underprivileged youth also came up in relation to the previously discussed topic around the preparation for the future, as can be seen in this response from Holly, a rural librarian from the central US, *"I think that to be successful in life and to break out of your cycle of poverty, they have to be able to really understand computers."*

Challenges of assessing computational thinking in libraries

What challenges do libraries face when assessing computational thinking?

Before discussing CT assessments specifically, we begin this section by briefly reviewing the approaches to assessment of library programming currently in use in libraries as reported by our 37 interviewees before shifting to the challenges they raised that are specific to trying to assess CT. Overall, the most common approaches to assessing library programming are taking attendance and tracking retention across sessions. The motivation for this approach was put succinctly by Brad (suburban Northeastern US), who said: *"it's successful if we get kids there"*. In general, this form of assessment is not problematic for CT-related library programming but provides little insight into the goals cited above such as attitudinal or learning outcomes. Beyond these numbers, other approaches for assessment referenced in our interviews included observations, surveys or questionnaires, impromptu interviews, and anecdotal feedback. Some of these approaches can provide insight directly related to the goals of CT programming, especially when incorporating existing CT-related surveys. For example, Caroline, a librarian from the rural Western US, said *"we did a program with NCWIT... they actually have a pre- and post-survey particularly looking at attitudes reviews"*, but, as will be discussed below, there are challenges associated with conducting this type of assessment in libraries.

The interview protocol used in this study included a number of opportunities for interviewees to voice the challenges they face in trying to assess their CT programming. This included questions such as "what challenges or constraints do you face with your CT programming?" and "how do you know if your program is successful"? One of the central challenges of assessing in libraries is rooted in the structure of the learning experience and the explicit recognition that libraries are not schools, so while learning is often a goal, assessment cannot feel like a school test. This can be seen in how Jenna, a rural librarian from the Northeast US, responded to a question about assessment *"Because we're a library ... our goal is lifelong learning. Whereas if I were a school, I might be more concerned that, did a particular skill get learned? Did a particular concept get understood? So I don't approach our evaluation in the way I would if I were in a school."* A former school librarian turned public librarian (Nicole, suburban Western US), also identified this difference in goals as it relates to learning *"If I were still in the like, 'I must teach these kids' mentality, then I would love it if they could get certain competencies checked off. But I also recognize that as a public library, that isn't my job"*. This difference has implications when it comes to the question of assessment. Whereas schools focus on content mastery and skill development, as stated above, libraries end up relying on less intrusive measures of success (e.g., attendance or retention) that provide little in the way of insight into measuring desired outcomes such as engagement or interest development. The result of this difference is that CT assessments developed for formal contexts may not be suitable for informal settings like libraries.

Other library staff identified CT assessment challenges associated with the characteristics of CT content and how it does not easily align with existing assessments, the way that mathematics or reading proficiency might. For example, as Matt, a rural librarian from the Northeastern US, said, *"I think there's big picture assessment with computational thinking. You got to understand where you start and where you're headed, finding out the steps that you need to get in between and breaking that down into not just what a computer would do, little bits and pieces at a time, to put those together, and then creating a toolbox where you have different tools and different pieces of logic and different pieces of computational thinking you can assemble together"*. It is not immediately obvious what an assessment for *"understanding where you are heading"* or *"assembling different pieces of logic"* would look like. This relational concern can also be seen in quotes from library staff related to the ways that CT intersects with other content areas and learning objectives, e.g., *"So if we could, making the connection in between computational thinking and social-emotional learning, that would be something that would be valuable because I think there are connections"* (Alice, urban Midwestern US).

The newness of CT and CT-related technologies was also raised as a source of challenges for assessment in libraries. This can be seen in comments related to the concepts and technologies themselves, e.g., *"Like a 3D printing program, the concept of 3D printing is often new to everyone in the room regardless of their age"* (Jenna, rural Northeastern US). This same concern was raised with respect to those currently working in libraries, as Brittany, a rural librarian from the Southwestern US, stated, *"I think the biggest challenge sometimes is just mine and my staff's own knowledge ... especially coding or just really understanding different types of programs that could utilize computational thinking"*. A second librarian (Matt, rural Northeastern US) echoed a similar experience, saying *"Most of what we're doing is being implemented by people who have very little to no computational thinking experience"*. This is not necessarily always a negative, as Jean, a suburban librarian from Northeastern US, pointed out, *"We're teaching ourselves just as much as we're teaching them, which honestly is sometimes better because then they get the luxury of being able to teach us, which always is very empowering when you're able to teach an adult something that they didn't know. So I think the kids really enjoy that. We're all learning it together."* This has direct implications for assessment, as in the absence of existing CT assessment specifically designed for informal environments (as stated above), it is then left to the library staff themselves to develop assessment, but the general lack of prior experience with the content makes this particularly challenging.

Logistical challenges of assessments were also raised by the library staff we interviewed. For example, Sandra, a rural librarian from the Midwestern US, responded "*I have a really hard time getting them to take a survey. I'm about ready to start bribing them with candy. It's hard to get them to do that and I think part of it is, is that I tend to hand them the survey about the time they need to leave, so that's hard. But I can't really give them the survey in the middle of things, before I'm all done, can I?*" This struggle also relates to the earlier discussion of the tension between assessment and informal settings where youth are choosing to participate, so engagement and fun are important considerations.

To summarize, the primary challenges to conducting assessments voiced in the interviews were the lack of clarity around the nature of CT or how to assess it, tensions between assessment and the informal nature of libraries; not having the CT knowledge or experience to develop effective assessments, and general logistical challenges associated with assessing a relatively new content area. Having reviewed the motivations for bringing CT programming into libraries and the challenges of assessing it, we now turn towards questions that gave the interviewees the opportunity to talk about what they wish they could do when it comes to assessing CT in their libraries.

Aspirational computational thinking assessments

How would public library staff like to assess computational thinking?

As part of the interview protocol, library staff members were asked how they would like to assess their CT programs. In asking this aspirational CT assessment question, we gain insight into where libraries are hoping to go with the assessment of their CT programs and shed light on the gap in assessment that currently exists.

The most common response given to what libraries would like to assess in terms of CT can be described as measuring long-term impact. As Caroline (rural Western US) succinctly put it: "*Assessment isn't about what happened in six weeks. It's what happens over years and years.*" As for what longitudinal assessment might look like, Brittany from the rural Southwestern US said: "*It would be really neat to be able to follow these kids through high school to see if the fact that they were involved in a coding club outside of their formal education in school if that made any difference*". A second interviewee, Lisa, a librarian from the rural Midwestern US, mentioned how she would like to see their patrons become mentors for future programs: "*I think if they were to start coming in and if we could grow them into teen volunteers who came back to help mentor the younger kids, that would be one thing, but I don't know how you express that.*"

A second aspirational goal of CT assessment mentioned by interviewees was to measure gains in knowledge of CT content and how the library programming helped students learn CT. As stated by Jeremy from an urban library in the Western US: "*Like before you came to the program, what was your understanding of these concepts and practices about computational thinking and how did that change at the end of the program?*" This goal for CT assessments can also be seen in Cole (suburban Midwest), who rhetorically asked: "*can [the child] actually demonstrate that there was an understanding of a concept or some comprehension or as I mentioned before, like some vocabulary that they picked up that they understand?*" In these responses, we see rather school-like views of assessment serving as a means to measure content mastery. However, as Lila, a librarian from the suburban Southeastern US, notes, this can be challenging given the amorphous nature of CT, saying *"I feel like some of my goals are so abstract that I'm not sure the kids could really viably tell me."*

Another form of assessment library staff were interested in was not focused on content mastery or learning, but rather gaining insight into youth's areas of interest and perceptions of the CT programs currently run, or, as Jean from the suburban Northeast puts it, *"seeing what the kids actually want to be doing"*. The goals of such assessments were stated to be both as a way to understand the learning experience but also to serve as a way to inform future program design, as Lauren, a librarian from the suburban Northeast US, said, she would like to ask the youth that visit her library: *"what was your favorite thing about the lesson? What are you still confused on? What would you like to know in the future? Like that kind of thing, which would help me kind of build the next classes."* This same sentiment can be found in Holly's response where she says that she is looking for a *"kind of metric [that] will tell me if I'm hitting the right spot and help me design future programming"*. A third example of this can be seen with Julia, a librarian from the rural Southeastern US, who said she would like *"a kind of a cumulative evaluation at the end of the programs to see just anonymously what they thought. How did, did you, did this work for you? Did your children enjoy this? What would you like to see in the future?"* This comment is interesting as it seeks the input of both the kids and their parents for assessing the program. While there are surveys that exist to assess attitudes towards computing that could be useful for shedding light on this desired information (e.g., the Computing Attitudes Survey (Dorn & Elliott Tew, 2015)), they have primarily been designed for formal learning contexts (or other contexts where youth can be required to complete them, like computing camps) so do not fit well in the free-choice setting of a library.

Library staff also articulated a desire for developmentally appropriate assessments, *"I would like to, I would love for, to have something that was developed for elementary students, specifically learning computer science, in afterschool programs, specifically or even generally"* (Alice, urban Midwest). This need was raised due to the fact that libraries support a range of ages and will have youth at different ability levels participating in the same activity, meaning a one-size-fits-all assessment might not be appropriate.

Finally, it is worth noting that several participants did not have an answer to this question, suggesting they do not spend much time thinking about assessments or prioritizing them in their work. This was succinctly captured in how Matt, a librarian from the rural Northeastern US, responded to our aspirational CT assessment question: *"I'm open to any ideas, but there's nothing that I have a burning desire to do as far as measurement goes"*. We include this not to disparage the library staff but to shed light on the emphasis some library staff place on assessment and measurement relative to the large set of responsibilities, goals, and challenges they face in their jobs.

To summarize, aspirational CT assessments shared by the library staff included assessments to measure gains in knowledge of CT, a way to measure the long-term impacts of CT programming in libraries, developmental- and age-appropriate assessments, and instruments to gauge interest in CT programming in libraries.

Discussion

Alignment between goals and desired assessments

In analyzing the responses given by library staff to questions related to the goals and motivations for their CT programming and mapping them on to the types of assessments that they wish were available, we find general alignment between the two. Library staff cited conceptual learning gains, positive attitudes, and experiences towards CT, and preparation for future learning and careers as key motivators for bringing CT into their libraries. At the same time, when asked what they would like to assess, we find a similar list: content mastery, interest development, and measuring long-term impacts of participation in CT programming (i.e. impacts on future school coursework and career choices). This presents a consistent vision for the purpose of CT programming in libraries and a clear alignment between what library staff want to do and what they want to measure in their programs.

However, in reviewing the assessment approaches currently in use and what is being measured, we find a disconnect between these aforementioned goals and the assessment strategies currently in use. Library staff reported mostly relying on attendance and retention numbers to assess programs. These numbers may serve as a proxy for demonstrating interest, but they provide little insight into determining whether the CT programming was successful in achieving these goals (e.g., conceptual learning, positive perception towards CT). Other cited assessment strategies currently in use, such as observations or impromptu interviews, are closer to the types of aspirational CT assessments mentioned (e.g., greater insight into emerging interest) but these current strategies in practice lack the systematicity or structure (e.g., interviews or observation protocols) that are important for providing a clear and consistent understanding of the impact of CT programming.

It is also worth mentioning that one of the explicitly stated goals for bringing CT into libraries, that of providing access to tools and technology that youth may otherwise not be able to experience, was not mentioned in response to the interview questions around desired assessments. Whereas most of the other motivations speak to some sort of change in the program attendees over time (i.e. learning a new skill, developing a new interest), the goal of physical access does not have a temporal dimension, but instead captures in-the-moment events (i.e. a youth using a technology they otherwise may not have had). In this way, the assessment of access need not be administered more than once (i.e. is not longitudinal) and can be captured through the currently used strategies at the library (e.g., attendance, observation, impromptu interviews), which means it may be an easier metric to capture as it is not subject to the challenges related to retention and structure of library programming that many of the other dimensions rely on and can be challenging for libraries.

Difficulties in assessing some stated goals

One of the findings of this study is the inherent challenge of assessing some of the stated goals for CT programming in libraries. For example, if the goal of a library program is to better prepare youth to pursue technology-related careers, how do you know if the program was successful in achieving this? This is not just a question of timeline and logistics (i.e. there being a potential decade-long year gap between the CT program and the youth settling into a career) but more fundamental questions of causality and measure of impact. The various experiences and factors that feed into the decisions of an academic major or what career to pursue are complex to be able to measure the impact of something like participating in a library program in a meaningful way. The point here is not that the stated goals are unimportant or that library CT programming cannot have the desired impact but rather a recognition of the current limitations of assessment as it relates to such goals of library programming.

A second difficulty around assessing some of the stated CT goals is not nearly as intractable as the first but nevertheless one that needs to be recognized. The recency of the emergence of CT and the introduction of CT programming in libraries, coupled with the continued uncertainty around exactly what constitutes CT is a source of difficulty in assessing CT. This difficulty goes beyond the potential lack of expertise within a specific library but a recognition that we as a community of researchers and scholars do not yet know how to effectively assess some of the central tenets of CT. For example, how do you assess a person's mastery of the use of abstraction? Or their ability to effectively decompose problems or create modularized solutions that are more easily reused? Likewise, practices such as debugging and effective use of trial-and-error are central tools in the CT toolbelt but similarly lack reliable, valid assessments. We raise this point so as to align the challenges faced by libraries and library staff around the challenges of assessing CT with those of other educators and education researchers. In aligning these groups, we hope to highlight ways that advances made around CT assessment in one context may also be applied in the other. We see this as bidirectional, innovations from the informal learning communities may advance assessing CT in formal contexts and vise versa. Linking these two communities may serve as one way to help address some of the difficulties identified by the library staff interviewed in this study.

Assessing computational thinking in libraries versus schools

Throughout the interviews, the library staff we interviewed often discussed their relationship with schools. At times, this was in order to highlight similarities in the form of goals as it relates to CT and potential ways the two can complement each other. At other times, schools were brought up as a contrast to libraries and the nature of learning in library contexts. In particular, differences were brought up with respect to slightly different goals between the two, and the differences as it relates to the different forms assessment can (and should) take.

For example, several library staff pointed out how enjoyment and engagement are central goals of CT programming in libraries in a way that need not be a central concern for teachers in schools where attendance is compulsory. This has direct implications for assessment. If the assessments being administered in the library are "school-like" (i.e. pen-and-paper surveys), then youth will not take them, or will not take them seriously, rendering the data useless. This means an alternative form of assessment needs to be used, where the assessment itself is either engaging or is embedded within the activity itself. This is a growing area of interest in CT and there has been some recent work focused explicitly on formative CT assessment that prioritizes playfulness and engagement, such as the Beats Empire project (Basu et al., 2020).

Another interesting distinction raised about the differences between schools and libraries that has direct implications on the nature of assessment relates to the stated goals. Whereas schools largely prioritize conceptual learning and skill development, library programming also has an emphasis on interest development and developing other "soft skills" (e.g., communication, collaboration, perseverance). As a result, assessments designed for school contexts that reflect the priorities of schools (i.e. conceptual understanding) may not be a good fit for libraries given this difference in goals (along with the fact that pen-and-paper assessments are a poor fit for the informal nature of libraries). This is consequential as many of the assessments that currently exist, especially the subset that have been validated, have been created with school contexts in mind, rather than informal or interest-driven learning that libraries strive for.

Audiences and implications

This research has two primary audiences with different implications for each. The first audience is individuals working in or with libraries to offer CT programming. This work can provide a framework for reflecting and rethinking current and future CT programming, the goals of such programming, and potential ways to rethink how, why, and when the programming is being offered. The implications of this work to those audiences is to help those doing this work situated their motivations, challenges, and assessment goals relative to others in a similar position and help provide a language for future discussion and orient future efforts to bring CT into their libraries. The second is those seeking to support the creation of tools, curricular activities, and assessments for libraries. We commenced this work by capturing the voices of public library staff who are responsible for bringing CT into libraries to answer these research questions. In doing so, we synthesize the needs of library staff so as to inform those seeking to help them, in particular, this work provides a road map for future work, specifically as it relates to CT assessments for libraries.

Situating the contribution of this work within the literature review presented at its outset, this research advances our understanding of libraries as a context for engaging youth with CT broadly while specifically contributing to our understanding of challenges and opportunities related to assessing CT programming in libraries. This work bridges existing scholarship on the state of CT in libraries (Subramaniam et al., 2018b; Taylor et al., 2018) with the growing body of research on how to assess CT (Tang et al., 2020; other articles in this special issue). In focusing on assessment, it adds another dimension to the discussion that up to this point has been more focused on questions related to how to bring CT in libraries (e.g., Braun & Visser, 2017). Broadening the literature to include questions of assessment can help inform and improve the important work already done. A central finding of this research is highlighting the disconnect between the motivations library staff have for bringing CT into their libraries (e.g., positive CT experiences, conceptual CT learning, career readiness, access to technology) with the forms of assessment most frequently employed (e.g., attendance and retention measures, surveys). This disconnect is partially explained by the various challenges cited in bringing CT into libraries (e.g., informal nature of library programming, difficult to measure outcomes, lack of existing assessment tools). In identifying this disconnect, and further articulating the types of assessment library staff would like to use to evaluate the CT programming (e.g., learning gains, increased interest, long-term impact), this work both contributes to our understanding of the current state of assessing CT in libraries while also laying out a roadmap for future work.

A roadmap for future work

One of the goals of this work is to understand the current state of CT assessments in library contexts so as to set a research agenda for future work. In asking library staff questions related to the goals of their CT programming as well as their hopes for aspirational CT assessments and the challenges they currently face in assessing their programming, we gain insight into exactly what the needs are for a community of CT educators faced with meaningful, pertinent, and pressing assessment challenges. The findings from this work reveal an immediate need for more CT assessments that can be employed in libraries that can capture aspects including interest development, shifting perspectives towards computing and CT, and conceptual learning. At the same time, these assessments must adhere to the norms and constraints of library contexts. This includes the informal, opt-in nature of programming in libraries and a recognition that the person who may be administering the assessments might be as new to the subject area as the youth being assessed. While these may be significant challenges, so too are the potential benefits of succeeding in making progress towards useful CT assessments for libraries given the critical role they play in bringing CT learning opportunities to communities around the world. Future work of this research team will involve designing, developing, and testing such assessments in libraries that are actively offering various types of CT programming through a participatory approach (Subramaniam, 2016; Yip et al, 2019). As CT and its related concepts and practices become increasingly prevalent, and the role of libraries continues to grow, having accurate, informative, and adaptable CT assessment can go a long way in ensuring the CT programming happening in libraries around the world achieves its full potential.

Conclusion

The emergence of CT as an educational movement has resulted in a renewed focus on creating pathways for youth to meaningfully engage with technology and the powerful computational ideas that enable it. Increasingly, libraries are playing a significant role in providing CT learning opportunities to youth. As CT learning programs in libraries grow and mature, it is important that the people running these activities have the ability to assess whether or not the programming is achieving its goals. With this work, we focused on libraries as contexts for CT engagement and investigated what the goals of CT programming are in libraries and what challenges libraries face in trying to assess their current CT programming. We also presented findings as to what libraries would like to be able to assess. In doing so, we advance our understanding of the current state of CT programming in libraries and provide insight into why and how libraries are providing CT programming. This research contributes to the literature seeking to understand the various ways we can deliver on the goal of bringing CT to all youth and lays the groundwork for future research seeking to equip library staff with the tools they need to provide effective, engaging, and successful CT programming to the youth they serve.

Notes

1. We use the terms "programming" and "programs" in this paper to describe structured activities offered in libraries (e.g., "library programs" or "CT programming"). This is a common vernacular in library scholarship.
2. All names are pseudonyms.

Disclosure statement

No potential conflict of interest was reported by the authors.

Funding

This work was supported by the Institute of Museum and Library Services under Grant # LG-14-19-0079-19.

ORCID

David Weintrop http://orcid.org/0000-0002-3009-3899

References

American Library Association. (2019). *State of America's libraries 2019. Public Libraries*. News and Press Center. http://www.ala.org/news/state-americas-libraries-report-2019/public-libraries

American Library Association. (2020). *State of America's libraries 2020. Issues and trends*. News and Press Center. http://www.ala.org/news/state-americas-libraries-report-2020/issues-trends

Arastoopour Irgens, G., Dabholkar, S., Bain, C., Woods, P., Hall, K., Swanson, H., ... Wilensky, U. (2020). Modeling and measuring high school students' computational thinking practices in science. *Journal of Science Education and Technology*, *29*(1), 137–161. https://doi.org/10.1007/s10956-020-09811-1

Basu, S., Disalvo, B., Rutstein, D., Xu, Y., Roschelle, J., & Holbert, N. (2020, February). The role of evidence centered design and participatory design in a playful assessment for computational thinking about data. In *Proceedings of the 51st ACM Technical Symposium on Computer Science Education* (pp. 985–991). https://doi.org/10.1145/3328778.3366881

Berland, M., & Lee, V. R. (2011). Collaborative strategic board games as a site for distributed computational thinking. *International Journal of Game-Based Learning*, *1*(2), 65–81. https://doi.org/10.4018/ijgbl.2011040105

Boyatzis, R. E. (1998). *Transforming qualitative information: Thematic analysis and code development*. Sage Publications.

Braun, L., & Visser, M. (2017). *Ready to Code: Connecting youth to CS opportunities through libraries*. The American Library Association's Office for Information Technology Policy.

Braun, L. W., Hartman, M. L., Hughes-Hassell, S., Kumasi, K., & Yoke, B. (2014). The future of library services for and with teens: A call to action.

Brennan, K., & Resnick, M. (2012, April). New frameworks for studying and assessing the development of computational thinking. In *Proceedings of the 2012 annual meeting of the American educational research association, Vancouver, Canada* (Vol. 1, p. 25).

Cole, N., Walter, V., & Mitnick, E. (2013). *Outcomes + outreach the california summer reading outcomes initiative* »The Public Library Association. http://publiclibrariesonline.org/2013/05/outcomes/

Curzon, B. (2019). Computational thinking. In S. A. Fincher & A. V. Robins (Eds.), *The Cambridge handbook of computing education research* (pp. 513–546). Cambridge University Press.

Davis, K., Subramaniam, M., Hoffman, K. M., & Romeijn-Stout, M. (2018). *Technology use in rural and urban public libraries: Implication for connected learning in youth programming*. Connected Learning Summit.

Denner, J., Werner, L., & Ortiz, E. (2012). Computer games created by middle school girls: Can they be used to measure understanding of computer science concepts? *Computers & Education*, *58*(1), 240–249. https://doi.org/10.1016/j.compedu.2011.08.006

Dorn, B., & Elliott Tew, A. (2015). Empirical validation and application of the computing attitudes survey. *Computer Science Education*, *25*(1), 1–36. https://doi.org/10.1080/08993408.2015.1014142

Garmer, A. K. (2014). *Rising to the challenge: Re-envisioning public libraries*. Aspen Institute. https://csreports.aspeninstitute.org/documents/Aspen-LibrariesReport-2017-FINAL.pdf

Grover, S., & Pea, R. (2013). Computational thinking in K–12: A review of the state of the field. *Educational Researcher*, *42*(1), 38–43. https://doi.org/10.3102/0013189X12463051

Grover, S., Pea, R., & Cooper, S. (2015). Designing for deeper learning in a blended computer science course for middle school students. *Computer Science Education*, *25*(2), 199–237. https://doi.org/10.1080/08993408.2015.1033142

Guidara, M. (2018). Rethinking computational thinking for public libraries' youth programs: challenges and recommendations. *Pennsylvania Libraries: Research & Practice*, *6*(2), 75–85. https://doi.org/10.5195/palrap.2018.188

Hoffman, K. M., Subramaniam, M., Kawas, S., Scaff, L., & Davis, K. (2016).Connected libraries: Surveying the current landscape and charting a path to the future [White paper]. The ConnectedLib Project.

IFLA Library Map of the World. (2020). Retrieved September 17, 2020, from https://librarymap.ifla.org/

IMLS. (2019). *Public libraries in the United Statesfiscal year 2016*. The Institute.
Lee, V. (2019). Libraries will be essential to the smart and connected communities of the future. In V. Lee & A. Phillips (Eds.), *Reconceptualizing libraries: Opportunities from the learning and information sciences*. (pp. 9–16). Routledge.
Lee, V. R., & Recker, M. (2018). Paper circuits: A tangible, low threshold, low cost entry to computational thinking. *TechTrends, 62*(2), 197–203. https://doi.org/10.1007/s11528-017-0248-3
Martin, C. (2017). Libraries as facilitators of coding for all. *Knowledge Quest, 45*(3), 46–53. http://files.eric.ed.gov/fulltext/EJ1125376.pdf
Mouza, C., Marzocchi, A., Pan, Y.-C., & Pollock, L. (2016). Development, implementation, and outcomes of an equitable computer science after-school program: findings from middle-school students. *Journal of Research on Technology in Education, 48*(2), 84–104. https://doi.org/10.1080/15391523.2016.1146561
Penuel, W. R., Widman, S., Wortman, A., Chang-Order, J., Repko Erwin, M., Harris, M., & Van Horne, K. (2019). *Rockwood makerspace at multnomah county library*. Connected Learning Alliance.
Phillips, A., Lee, V., & Recker, M. (2019). Small town librarians as experienced engineers. In V. Lee & A. Phillips (Eds.), *Reconceptualizing libraries: Opportunities from the learning and information sciences*. (pp. 158–169). Routledge.
Prato, S. (2017). Beyond the computer age: A best practices intro for implementing library coding programs. *Children and Libraries, 15*(1), 19–21. https://doi.org/10.5860/cal.15n1.19
Pugnali, A., Sullivan, A., & Umashi Bers, M. (2017). The impact of user interface on young children's computational thinking. *Journal of Information Technology Education: Innovations in Practice, 16* (1), 171–193. https://www.learntechlib.org/p/180753/
Rose, R.N. (2015). *Survey of chief officers of state library agencies*. American Library Association
Smagorinsky, P. (2008). The method section as conceptual epicenter in constructing social science research reports. *Written Communication, 25*(3), 389–411. https://doi.org/10.1177/0741088308317815
Subramaniam, M. (2016). Designing the library of the future for and with teens: Librarians as the "connector" in connected learning. *The Journal of Research on Libraries and Young Adults, 7*(2), 1–18. http://www.yalsa.ala.org/jrlya/wp-content/uploads/2011/02/Subramaniam_Designing-the-Library_Final.pdf
Subramaniam, M., Johnston, M., Taylor, N. G., Moore, J., Magee, R., Drouillard, C., & Sanchez, J. (2018b). Expanding LIS Youth Services to embed computational thinking. *Association for Library and Information Science Education (ALISE) Conference*, Denver, Colorado.
Subramaniam, M., Kodama, C., Baylen, D. M., Burton, M., Fabicon, J. K., Hincks, K., Moniz, R., Smith, D., & Visser, M. (2019). *Computational thinking in libraries: Case studies of youth programs in action*. The American Library Association's Office for Information Technology Policy.
Subramaniam, M., Scaff, L., Kawas, S., Hoffman, K. M., & Davis, K. (2018a). Using technology to support equity and inclusion in youth library programming: Current practices and future opportunities. *The Library Quarterly, 88*(4), 315–331. https://doi.org/10.1086/699267
Tang, X., Yin, Y., Lin, Q., Hadad, R., & Zhai, X. (2020). Assessing computational thinking: A systematic review of empirical studies. *Computers & Education, 148(*April), 103798. https://doi.org/10.1016/j.compedu.2019.103798
Taylor, N. G., Moore, J., Visser, M., & Drouillard, C. (2018). Incorporating computational thinking into library graduate course goals and objectives. *School Library Research, 21*. http://www.ala.org/aasl/sites/ala.org.aasl/files/content/aaslpubsandjournals/slr/vol21/SLR_IncorporatingComputationalThinking_V21.pdf
Thomas, J. O., Rankin, Y., Minor, R., & Sun, L. (2017). Exploring the difficulties African-American middle school girls face enacting computational algorithmic thinking over three years while designing games for social change. *Computer Supported Cooperative Work (CSCW), 26*(4–6), 389–421. https://doi.org/10.1007/s10606-017-9292-y
Wang, J., Hong, H., Ravitz, J., & Hejazi Moghadam, S. (2016). Landscape of K-12 computer science education in the US: Perceptions, access, and barriers. In *Proceedings of the 47th ACM Technical Symposium on Computing Science Education - SIGCSE '16* (pp. 645–650). https://doi.org/10.1145/2839509.2844628

Weintrop, D., Beheshti, E., Horn, M. S., Orton, K., Jona, K., Trouille, L., & Wilensky, U. (2014) Interactive assessment tools for computational thinking in high school STEM classrooms. INTETAIN 2014, Chicago, IL.

Wing, D., & Meyers, E. (2014). Easy as pi: Designing a library program to support computational thinking in preteens. *BCLA Browser: Linking the Library Landscape*, *6*(3). https://www.bclabrowser.ca/index.php/browser/article/download/632/803

Wing, J. (2006). Computational thinking. *Communications of the ACM*, *45*(3), 33–35. https://doi.org/10.1145/1118178.1118215

Wu, M. L. (2018). Educational game design as gateway for operationalizing computational thinking skills among middle school students. *International Education Studies*, *11*(4), 15. https://doi.org/10.5539/ies.v11n4p15

Yip, J. C., Lee, K. J., & Lee, J. H. (2019). Design partnerships for participatory librarianship: A conceptual model for understanding librarians co-designing with digital youth. *Journal of the Association for Information Science and Technology*, *71*(10). https://doi.org/10.1002/asi.24320

Young Adult Library Services Association. (2016). *Future ready with the library: Connecting with the communities for college and career readiness services*. http://www.ala.org/yalsa/future-ready-library

Young Adult Library Services Association. (2018). *Transforming teen services: A train the trainer approach*. http://www.ala.org/yalsa/train-trainer-project

Index

Note: **Bold** page numbers refer to tables; *Italic* page numbers refer to figures and page numbers followed by "n" denote endnotes.